BAKED TO PERFECTION

KATARINA CERMELJ

BAKED TO PERFECTION

DELICIOUS GLUTEN-FREE RECIPES WITH A PINCH OF SCIENCE

BLOOMSBURY PUBLISHING
LONDON • OXFORD • NEW YORK • NEW DELHI • SYDNEY

FOR MY PARENTS.

Thank you for your love and support and endless patience
in dealing with all the food (and the dirty dishes),
the let-me-test-this-just-one-more-time experiments
and the overthinking that went into creating this book.
Love you so much.

BLOOMSBURY PUBLISHING
Bloomsbury Publishing Plc
50 Bedford Square, London, WC1B 3DP, UK
29 Earlsfort Terrace, Dublin 2, Ireland

BLOOMSBURY, BLOOMSBURY PUBLISHING and the Diana logo are trademarks
of Bloomsbury Publishing Plc.

First published in Great Britain 2021

A catalogue record for this book is available from the British Library.

Library of Congress Cataloguing-in-Publication data has been applied for.

ISBN: HB: 978-1-5266-1348-6; eBook: 978-1-5266-1347-9

10 9 8 7 6

Project Editor: Judy Barratt
Designer: Peter Dawson, Grade Design Consultants Ltd
Photographer: Katarina Cermelj
Illustrator: Katarina Cermelj
Indexer: Hilary Bird
Proofreader: Jo Murray

Printed and bound in Germany by Mohn Media

FSC
www.fsc.org

MIX
Paper from
responsible sources
FSC® C011124

To find out more about our authors and books visit www.bloomsbury.com and sign
up for our newsletters.

CONTENTS

INTRODUCTION

We're here to talk about the quantum mechanical analysis of the behaviour of sugar molecules when surrounded by particles of gluten-free flour. Wait, wait, wait, I'm kidding. Don't put the book down yet. Truly, the scariest things in this book are the heart-stopping, jaw-dropping, ultra-fudgy, caramel-stuffed brownies on page 136, simply because they're so delicious they're borderline dangerous.

Let's start this again.

This is a gluten-free cook book. Whether you've picked it up because you or someone close to you has a gluten intolerance, because you really, really loved the cover, or simply by accident, it matters not. Because this cook book and these recipes are for everyone.

My own gluten-free journey started – well, as most gluten-free journeys do. With feeling ill and miserable and then, partly by chance and partly by hours of research that involved reading through one-too-many online forums, coming to the conclusion that eliminating gluten might help. It did. And it also brought along the question of, what on Earth do I eat now?

I like smoothies and quinoa as much as the next person, but I also love chocolate cake. And brownies. And apple pie. And all the other delicious things that I've baked a hundred times before – with plain wheat flour.

So, unless I wanted to give up cookies and cake and brownies – oh my, what a dreadful thought that was – I needed to master the intricacies of gluten-free baking. Thankfully, I had an ace up my sleeve. That is, my knowledge of chemistry and an almost obsessive love of research.

You see, my gluten-free baking adventure started just as I was finishing my undergraduate Chemistry degree at the University of Oxford, and then continued through my PhD studies of Inorganic Chemistry. Therefore, it felt natural to approach the development of gluten-free recipes as I would a new research project in the laboratory.

What's more, my scientific background gave me a particular insight into the ingredients, their properties and their interactions with each other. I wasn't just mixing together gluten-free flour with butter, sugar and eggs, and hoping for the best. Instead, I was thinking about the wet:dry ingredients' ratio, the different moisture responses of the various gluten-free flours, and the speed of heat penetration as it varies with the shape and depth of the baking tin.

Pair my scientific knowledge with a heaping tablespoon of stubbornness and a generous pinch of nerdiness – and you've got yourself a mess of overcomplicated Excel sheets, endless notes and illegible scribbles at the sides of numerous recipes covered in suspicious blotches of what might have been (once upon a time) cupcake batter and chocolate buttercream.

More importantly, you've got yourself (over time and after prolonged systematic experimentation) gluten-free recipes that work and work well. Gluten-free recipes that are simple and give mouthwatering, drool-inducing results – in a reproducible manner. Because if there's one thing we scientists are all about, it's reproducibility.

Don't worry. You won't have to go through the whole process of experimentation. There are no three-page Excel sheets in this book. I've done all the experiments for you and condensed them into simple, reliable recipes that will make your own gluten-free life so much richer and infinitely more delicious.

But there's more! The recipes in this book don't just give you the list of ingredients and instructions. They also give you the evidence – the scientific explanations, if you will – for why the recipes work.

This isn't just to satisfy my own nerdiness. It's primarily to help you become a more skilful (and successful) gluten-free baker. Once you understand the why, you can easily afford to deviate from the recipe and adapt it to your own taste. Because here's the thing I love: baking isn't as rigid as many would have you believe. Yes, it is a science of sorts, but rather than making it restrictive, science actually makes baking very flexible, as long as you know which things to leave alone.

It could be a specific step (such as whisking the eggs and sugar into pale, fluffy submission) or a set of ingredients (such as the ratio of wet:dry ingredients), but as long as you don't change that, you can do pretty much anything else you want. Adding flavours, tweaking textures, substituting ingredients… it all becomes incredibly simple once you understand the why.

That's my goal with this book. Yes, I'm giving you tried-and-tested (and then tested again, and again) gluten-free recipes that work and work well; and I'm giving you an insight into the world of gluten-free baking that shows that 'gluten-free' doesn't equate to 'delicious-free', but I'm also giving you the knowledge to make those recipes your own and develop them further, beyond my ingredients and instructions.

LET'S TALK MORE ABOUT GLUTEN-FREE

Years after I started my gluten-free journey, gluten-free still has a bad reputation. At best, you'll find people claiming gluten-free can be 'just as good' as the 'real thing'. At worst, you'll find the words 'bland', 'dry' and 'cardboard' thrown around like sad, depressing confetti.

I disagree. On both counts. Because here's the thing: gluten-free isn't just as good as the real thing – gluten-free is the real thing. The luxurious, moist, indulgent gluten-free chocolate cake isn't delicious despite being gluten-free. It's delicious while being gluten-free. And that's all there is to it.

Of course, in order to master gluten-free baking, you first need to shake off the rules and notions of wheat-based baking. It's a different set of ingredients with a different set of properties – naturally, then, the rules are different too. And yet, we persist in trying to use recipes developed for wheat-based baking, and then, when the results are dry and claggy and not at all pleasant, we blame gluten-free.

Let me ask you this: if you attempted to make brownies by following a cookie recipe, and they turned out crispy instead of gooey, hard instead of fudgy… would you blame the recipe? Would you give up baking and brownies and chocolate for ever? Would you proclaim brownies a waste of time?

Of course not. (At least, I hope not.) You would wise up and use a brownie recipe next time. You would accept the fact that you're playing a different game and you would change the rules. The same is true with gluten-free baking.

SUPER-DELICIOUS CHOCOLATE CAKES... THAT ALSO HAPPEN TO BE GLUTEN-FREE

WHERE THE SCIENCE COMES IN

You've heard it said numerous times before, and you'll probably hear it a hundred times more: baking is a science. And gluten-free baking even more so. Rather than being restrictive, there is a particular freedom to be found in the scientific understanding.

I know that science isn't everyone's idea of a good time, and you might be having nightmare-ish flashbacks to school Chemistry lessons right now. But, before you despair, this isn't a dry encyclopedia of the exact science of gluten-free baking, complete with pages-long formulae and complex terminology. The science we're talking about here tells us how to achieve that elusive, shiny, paper-thin crust on brownies – every time and without having to rely on sheer dumb luck. Or why gluten-free baking usually requires a higher wet:dry ingredients' ratio. Or what is the 'reverse creaming method' and why it gives a more delicate sponge with a velvety crumb texture. It's the delicious kind of science. The kind that results in fudgy brownies, melt-in-the-mouth cookies and gluten-free bread that actually has the taste, texture and appearance of bread.

For best results, treat this cook book as the slightly nerdy friend you turn to when you get stuck and your gluten-free pie crust is too crumbly or too soggy. It has all the answers, right where you need them, presented in short, easily accessible blurbs.

You can find the science in three places in the book: in the next chapter, in the opening sections of each recipe chapter and as footnotes in individual recipes (I might have on occasion snuck it into the recipe descriptions as well). The science in this introduction serves as a gentle foray into the world of gluten-free baking and the general things you should keep in mind as you start making the recipes.

The science in the opening sections of each chapter focuses on ingredients, methods and other things specifically relevant to that kind of baking. Finally, the notes in the recipes explain and give details on the most important ingredients and techniques.

'THERE'S NO WAY THIS IS GLUTEN-FREE!'

Once you begin to understand the whys and hows (that is, the science) of gluten-free baking, prepare yourself to gain a new and constant companion – the exclamations of wonder from people who cannot possibly comprehend that your mouthwatering bakes truly are gluten-free. (Usually accompanied by shocked faces and followed by requests for *more, please and thank you.*)

What's even more exciting is that as you use the book, you'll become more confident and comfortable wielding the whisk, mixing the batter and adjusting the baking time. Straying from the well-defined path of a written recipe, you'll start making tweaks and scribbling notes about potential new flavour combinations and substitutions to try.

Once the science is clear, it's as though a whole new world of gluten-free baking stretches out before you. Luxurious cakes, dainty cookies, the gooiest of brownies, proper crusty bread, even gluten-free éclairs so beautiful like they've popped straight out of a French pâtisserie – nothing is impossible or beyond your skill. And if a bake fails (spectacularly and dismally, as it does and it will, no doubt about it), rather than seeing it as the end of your fledgling gluten-free baking adventure, it's just a new chance to learn and figure out what's going on in your bake both in and out of the oven. As it happens, the answer to your question probably already lies between these pages.

THE KIND OF RECIPES YOU'LL FIND IN THIS BOOK

1) Fundamental, basic recipes
The point of this gluten-free baking book is to function less as a collection of innovative, never-seen-before recipe ideas, and more as a go-to manual with all the fundamental, core gluten-free recipes from

which all other gluten-free bakes follow. The majority of recipes within these pages belong to this category.

That's not to say that these recipes aren't ground-breaking. They are. But in a subtler way than you might expect. They're ground-breaking because of the way in which they combine the ingredients, the techniques they use and the ingredient ratios, to result in gluten-free bakes the likes of which you might never have encountered before.

While the flavour combinations and the recipes themselves are more than likely old, familiar friends (like a chocolate cake or burger buns), the reasons behind using a specific ingredient, or a particular step or technique, might be less familiar.

At first, the world of gluten-free baking can seem overwhelming. And even if you're a far more experienced gluten-free baker looking only to refine your skills, you may find that going back to basics is the best way to do that. Now, don't get me wrong – yes, I do want to inspire you. But I want this inspiration to come more from an awareness of what's possible and achievable and less from a flavour combination you've never seen before.

And finally – while an elaborately decorated, three-tiered cake or a 15-component entremet are amazing and breathtaking, nothing (and I do mean nothing) beats a slice of perfectly baked chocolate cake or a piece of warm apple pie topped with a scoop of vanilla ice cream when it comes to comfort and that all-encompassing sense of happiness.

FAMILIAR SOFT DELICIOUSNESS… BUT PREPARED USING UNFAMILIAR METHODS

2) Recipe variations

I know I just spent half a page talking about how it's all about core recipes in here, but I do want you to go forth and experiment further in your own kitchen, creating your own unique masterpieces. Once you've mastered the basics, obviously.

To kick-start your sense of gluten-free-baking adventure, I've included my own suggestions – just gentle nudges – for twists and variations at the end of some recipes. For example, you could add tahini paste to the chocolate chip cookies on page 152 for an additional layer of flavour that pairs beautifully with the chocolate chips. Or, you might add raspberries to the white chocolate brownies on page 134 if you were looking for a perfect match that gives a fruity tang.

You see? Once you start thinking about it, the possibilities are endless. As you go through the recipes in the book, I want you to always think beyond the ingredients I've listed. And that's why understanding the science is so crucial to becoming an independent, innovative baker – once you know the why behind each and every ingredient and step, you know whether changing or replacing it will affect the bake. What's more, you'll know how it will affect the bake. And with that – nothing is impossible.

3) Recipes you can't live without

Okay, I realise that calling them recipes 'you can't live without' might be misleading – because, in fact, all the recipes in this book are so spectacular, you'll want to make them again and again. But this last category includes recipes that introduce only few (or no) new nuggets of knowledge, and yet I simply couldn't not include them in the book because of their sheer, heart-stopping deliciousness.

It's the marshmallow brownies, with the tantalising combination of a fudgy, chewy brownie and a fluffy cloud of toasted sweetness that is Swiss 'marshmallow' meringue (see page 142). It's the babka that has a very similar dough to the cinnamon rolls – but I simply couldn't bear not to share the beauty of the chocolate fudge filling swirling with brioche dough in the most delicious, drool-inducing dance ever to happen in my kitchen (page 308).

Depriving you of these recipes would be rather cruel – so I didn't. You're welcome.

THE RECIPES YOU WON'T FIND IN THIS BOOK

While it's incredibly important you know what kind of recipes I've included, it's probably even more important that you're aware what sort of recipes I've omitted and why.

Clearly, my reasoning for what's in and what's out is subjective and entirely my own. But I made those decisions in order to create enough room for the recipes and information that will most help you truly to master gluten-free baking.

1) Healthy recipes (or rather, recipes whose main claim is being healthy)

Listen, I'm all for healthy recipes and a balanced diet. But this book is not about that. If you're looking for a collection of recipes that are not only gluten, but also grain, sugar, dairy, egg and whateverelse free, this isn't a book for you. Too many times I've browsed the web and the bookstore for gluten-free-cake or -cookie recipe inspiration, only to encounter one 'healthy' recipe after another.

There's nothing wrong with that. But here, in this book, I want each and every bake you make to be decadent and indulgent, to warm your food-loving soul and to bring you that unique joy that only comfort food can.

That said, I am aware that certain starch-heavy, gluten-free flour blends are a cause of concern for some. With that in mind, one of my DIY gluten-free flour blend recipes (see page 19) isn't based on white rice and it contains significantly less starch than many store-bought brands (only 40% to a more usual 70% starchy flours). While you might think that I should eliminate starches altogether, in fact that would negatively impact the texture of most bakes. Moreover, many recipes in here are occasional treats and indulgences, and meant to be enjoyed as part of a balanced, soul-nourishing lifestyle. (And, in my humble opinion, such a lifestyle pretty much requires a cake or brownie every now and then.)

2) Naturally gluten-free recipes

I admit, finding naturally gluten-free recipes in gluten-free recipe collections is something of a pet peeve of mine, especially when it comes to collections that claim to teach you all the intricacies and how-tos involved in gluten-free cooking and baking. That's not to say I can't appreciate the chocolatey decadence of a flourless chocolate cake, or the comfort that comes along with a scoop of homemade ice cream.

To me, such recipes would take away valuable space from the very recipes that actually require our attention; recipes that might seem impossible to accomplish in the absence of gluten – fluffy cakes, chewy cookies, fudgy brownies, the perfect boiled and baked bagels. Why (in a book that's all about mastering gluten-free baking) would I waste space telling you about making crème brûlée, when I can use it to explain why the recipe for the gluten-free choux pastry on page 336 actually works?

Therefore, if a recipe is naturally gluten-free in that it doesn't contain any flour to begin with – no, you won't find it here.

3) Unnecessarily complicated recipes with inaccessible ingredients

We all know those recipes that seem almost to require you to grow your own apple tree, harvest the apples, cut them into precisely 1.3mm thin slices, douse them in a rare kind of sugar with an exact type of sugar crystal, drizzle them in 2⅕ tablespoons of lemon juice from lemons that can come from only one particular village in southern Italy, and then marinate this in the fridge for 3 days… to make a simple apple pie.

As a scientist at heart, I'm all about precision – but only insofar as it actually affects the final result in an important, can't-live-without-it way. So, when you attempt my recipes, do so knowing that when I tell you to do something or to use a particular ingredient, there's a reason behind it. And you should do it.

But because I don't want you merely to follow blindly my instructions, I've included notes that explain any unusual and/or important step or ingredient. All the ingredients should be easily available either in your local supermarket or online.

4) Hacks and shortcuts that compromise on flavour or texture

Just like I don't unnecessarily complicate recipes, I don't believe in simplifying them to the point where simplification negatively affects the bakes' flavour or texture. The same goes for the time it takes to make a recipe from start to finish. If, say, a bread dough will be easier to handle and have a better flavour after an hour chilling in the fridge, I won't go recommending you skip that step just to save time.

I know many of us are short on time and, believe me, I'm a very impatient baker. But for 99% of the recipes in this book, the active time (that is, the time you actually spend doing active work on the bake) is much, much shorter than the total time required to make a recipe. I've included the most relevant timings at the top of each recipe. First, there's the 'prep' time, which is the time you'll spend actively engaging with the bake (stirring, beating, whisking, assembling and so on). Then, there are the other timings (such as baking, chilling and rising) that are there to help you judge how long you can (and should) allow the bake to do its own thing. If you're feeling particularly organised, you can, of course, get on with other parts of the recipe during that time. For example, you can make the buttercream while your chocolate sponges are cooling; or prepare the tart filling while the pastry crust is blind baking. Some recipes also require a cooking step (like preparing a caramel or reducing the fruit juices when making a pie), which is summed up in the 'cook time'. Note that the 'cook time' doesn't include simple heating steps, such as heating the double cream for a ganache or the egg whites and sugar for Swiss meringue.

GLUTEN-FREE PERFECTION AWAITS YOU

It's a lofty claim – that all the bakes in this book have the potential to be perfect. Some might say I'm setting myself up for failure. But here's the thing: I **want** you to approach the bakes with high expectations. I don't want you to start your gluten-free journey thinking you'll be merely content or 'okay' with them. I want you to take your first gluten-free baking steps convinced you'll taste sheer and utter… perfection.

There's another reason I like to almost overuse the words 'perfect' and 'ideal' (they come up a lot). Any and all who know me 'in real life' know that I'm a perfectionist, and (when I'm not driving myself crazy trying to improve on an already improved recipe) rather proud of it, in a roundabout way. Almost every recipe in this book has gone through a minimum of ten variations, some through as many as thirty or forty. Some of these refinements were minuscule, so small many would think me insane. Does ¼ teaspoon of baking powder or 1g of xanthan gum really make that much difference? Does whisking the eggs with the sugar rather than simply stirring them together really give a drastically different brownie? Sometimes, it does. Other times, it doesn't. But it took making all those variations to determine the truly best incarnation of a bake – and the knowledge I've taken away from both the successes and the failures is even more valuable than all the recipes I've developed along the way. Talking about perfection in the context of this book, therefore, doesn't refer only to the quality of the recipes themselves, but also to the process of getting them to that point.

Of course, it would be incredibly easy to go down the rabbit hole of the question: what is perfection anyway? Especially when it comes to food, perfection is incredibly subjective. What I might find pleasing, you might find less so. That said, here's the wonderful thing about a cook book that gives not only the recipes but also the reasoning and the science behind them: yes, I've taken these recipes to my own point of perfection (one that I think you'll love too), but I'm also giving you the tools you'll need to achieve your own gluten-free perfection.

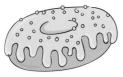

GLUTEN-FREE BAKING BASICS

WHAT IS GLUTEN?

Before we delve into the (not scary at all) science of gluten-free baking, it's important to understand what gluten actually is and the role it plays in wheat-based baking. Because at the end of the day, a large part of gluten-free baking is trying to recreate textures and structures that occur effortlessly and naturally in the presence of gluten.

The basic definition of gluten is 'a mixture of two proteins present in cereal grains, especially wheat, which is responsible for the elastic texture of dough'. The origin of the word dates back to the 16th century, to a Latin expression for 'glue'. These origins tell us much about the role that gluten plays in baked goods – it gives them the characteristic elasticity and essentially acts as a glue that keeps the bakes together.

Moreover, the gluten proteins determine other characteristic properties of wheat flour, such as its water-absorption capacity. That is, how much moisture or liquid it absorbs both before and during baking. When it comes to baked goods, the gluten proteins therefore determine how dry or moist, crumbly or 'springy' your bakes are.

Even in wheat-based baking, however, most bakers will have a love–hate relationship with gluten. While it's the most diligent helper when it comes to creating the characteristic crumb of a loaf of bread, it takes on the role of tricky opponent when making things like cookies and pie crust. For these, a baker needs to minimise the effects of gluten, lest he or she ends up with gummy cookies and a tough crust.

In gluten-free baking, the problems we face are exactly the opposite. Now, the cookies easily remain crisp (or chewy, as desired), the pie crust buttery and flaky, and the cake sponges delicate and with the perfect crumb. Bread, however? There, it gets trickier. That's not to say that baking proper gluten-free bread is an impossibility, it means only that we need to go about it in a different way.

GLUTEN-FREE FLOURS

The world of gluten-free flours can be overwhelming – there are just so many of them! In this book, I'll focus on my go-to ones, those I use daily in everything from cakes and cookies to pastry and bread. What follows is by no means an exhaustive list. Rather, it is really meant as a gentle introduction to gluten-free flours and their properties. Don't get me wrong: you could comfortably go on for years using only the flours that you'll encounter within these pages (with mouthwatering, eye-popping results) – but do also keep in mind that there are many, many flours beyond the ones I mention.

That said, the way we think about gluten-free flours, their properties and applications is universal – and once you understand the hows and whys, you'll be able to use pretty much any new gluten-free flour with no problem whatsoever.

Starch + protein flours

Gluten-free flours fall roughly into two groups: starch flours and protein flours. This distinction is primarily based on the protein content and the elasticity each flour gives to the bakes, and it also closely correlates to the way the flours interact with any liquid ingredients in the recipe (what I call their 'water-absorption capacity').

Broadly speaking, **protein gluten-free flours** (brown rice, buckwheat, maize, millet, oat, quinoa, sorghum and white teff flour) are the flavour- and structure-carriers of a gluten-free bake. They provide (a very small amount of) elasticity, preventing the bakes from being too crumbly. They also have a higher water-absorption capacity, which means that bakes with a higher protein-flour content are slower to dry out, all other things being equal.

Starch gluten-free flours (arrowroot starch, cornflour, potato starch, tapioca starch and white rice flour), on the other hand, are there to make the bakes softer, fluffier and airier. Because they have a smaller protein content and a lower water-absorption capacity, they make sure the cakes have a pleasant, open, melt-in-your-mouth crumb and the pastry isn't too gummy.

However, the real magic of gluten-free baking comes from choosing the right mixture of protein and starch flours that give you the right balance of elasticity, fluffiness and moistness and an open, airy crumb. A cake made from only starch flours will crumble away to nothing, whereas using exclusively protein flours might transform it into a dense, sticky brick. That's where gluten-free flour blends come into play – but more on that below. For now, let me just tell you that I've done the work for you, and figured out the perfect ratios of starch and protein flours that give the most delicious results across the board.

The distinction between the starch and protein flours is not the end of the story. Especially within the protein-flour group, there's a gradation of properties – particularly when it comes to the elasticity and the 'heaviness' they impart to the bakes.

'Additional' flours

To make things even more interesting (and, truth be told, confusing), there are additional flours that don't really fit in either of the two primary groups. I'm talking about almond flour (and other nut flours), coconut flour, chickpea flour and similar. While I'm not the biggest fan of chickpea flour because of its almost overwhelmingly strong flavour that isn't best suited to desserts and other sweet treats, I do use almond flour quite frequently, especially for cakes and cupcakes.

While they resemble starch flours in that they don't contribute to the elasticity of the bake, they also give the bakes a certain weight. This, in addition to their rather high water-absorption capacity, is quite reminiscent of a protein flour. These mixed properties mean that it's best to treat them as a separate group, which often acts as an add-in rather than a central component of a gluten-free bake.

Are you overwhelmed yet? To make your life a bit easier, I've summed up the main gluten-free flours in Table 1 (opposite). But to really gain a proper understanding of gluten-free flours, you'll have to use them. Feel their texture, see how they behave in baking. A good test of their properties is to mix about 1 tablespoon of the flour into ½–1 tablespoon of water. The amount of water the flour absorbs will tell you much about its water-absorption capacity, and the texture of the mixture will show you whether you're dealing with a protein or a starch flour. The former, when mixed with water, will form a 'dough' with varying degrees of elasticity (of course, nowhere near that of wheat flours), whereas the latter will form a runny paste.

In Table 1, you'll notice that I categorise brown rice flour as a protein flour, but white rice flour is in with the starches. White rice flour does contain slightly less protein than brown rice flour, but its protein content is still greater than that of, say, tapioca or potato starch. Nonetheless, when it comes to how it behaves in actual bakes, I've found that white rice flour is closer to starch flours than protein ones. So, while this is by no means a perfect classification, it works well in practice. It does mean, though, that you can't reliably substitute white rice flour for other starch flours (or vice versa) – more on that later.

The importance of finely ground flours

A final note on gluten-free flours: always make sure that you use finely ground gluten-free flours (or blends). They should have a fine, powdery texture – think regular wheat flour rather than polenta. Using a coarse flour inhibits its interaction with the rest of the ingredients and will lead to disappointing results.

Table 1: The most important gluten-free flours and their relative protein contents, water-absorption capacities, and assignments to the groups of starch, protein or 'additional' flours.

GLUTEN-FREE FLOUR	PROTEIN CONTENT [1]	WATER-ABSORPTION CAPACITY [2]	ASSIGNMENT
almond flour	very high (low elasticity)	medium	additional
arrowroot starch	low	low	starch
buckwheat flour	high	high	protein
coconut flour	high (low elasticity)	very high	additional
cornflour	low	low	starch
maize flour	medium	high	protein
millet flour	medium–high	medium	protein
oat flour	high	high	protein
potato starch	low	low	starch
quinoa flour	high	high	protein
rice flour, brown	medium	medium	protein
rice flour, white	medium–low	medium	starch
sorghum flour	medium–high	high	protein
tapioca starch	low	low	starch
teff flour, white	medium–high	high	protein

[1] Note that while a higher protein content does contribute some elasticity to the bake, this elasticity is defined within the realm of gluten-free flours and is nearly negligible when compared to that of wheat flour.
[2] The amount of water absorbed by dry powders at room temperature (tested by mixing together 10g flour with 10g room-temperature water).

THE BINDERS: XANTHAN GUM + PSYLLIUM HUSK

Of course, we can't talk about gluten-free baking without mentioning xanthan gum and psyllium husk ('the binders'). Their basic function is to act as a gluten substitute, giving gluten-free bakes their elasticity and flexibility. This is especially important when using a blend with a low protein content, which could result in dry and crumbly results. Just a pinch of xanthan gum can completely change the bake, giving it that 'I-can't-believe-this-is-gluten-free!' quality.

Xanthan gum is a polysaccharide (all that means is that its chemical structure comprises several smaller sugar units) produced by the fermentation of glucose and sucrose with the help of special bacteria. I know this might sound a bit odd – but xanthan gum has been shown to be perfectly safe as a food additive and it's pretty much a miracle ingredient in gluten-free recipes.

In baking, we're primarily interested in the ability of xanthan gum to drastically increase the viscosity (that is, the thickness and apparent stickiness) of a liquid, even when present in low concentrations. If you add xanthan gum to water, you'll notice it forms a gel within seconds – depending on the relative amount of xanthan and water, this gel can be either loose or quite firm, but it always has a degree of elasticity to it. So, only a pinch of xanthan gum is sufficient to give a gluten-free dough or batter a stickiness and flexibility reminiscent of that achieved by gluten.

Finally, let me bust a myth about xanthan gum: because it is sometimes used to thicken sauces or soups in a way similar to cornflour, a rumour has been going around the Internet that you can replace xanthan gum with cornflour in gluten-free baking. That's. Not. True. Let me repeat: that's not true. You can't replace xanthan gum with cornflour (well, not if you want good results). While you can use either as a thickening agent, only xanthan gum acts as a gluten substitute.

Psyllium husk is a less processed additive than xanthan gum but performs a similar function in gluten-free baking. Unlike xanthan gum, however, psyllium husk is not a pure substance but rather comprises several compounds. The compounds we're interested in for our gluten-free bakes are a group of gel-forming long chains of sugar units, which behave in a manner similar to xanthan gum (as you'd expect given their similar chemical structures). And, as with xanthan gum, mixing psyllium husk with water makes a wobbly, somewhat elastic gel.

While the two binders do carry out similar functions, I've found that they are best suited to different applications. Xanthan gum works wonders for cakes, cookies, cupcakes and muffins, brownies and pastry – that is, pretty much all bakes apart from bread. When it comes to bread, psyllium husk is the star of the show, making the dough easier to handle and more closely resemble wheat-based bread dough. As you browse through the recipes, you'll notice that psyllium husk is really restricted to only the bread chapter – but there, it makes all the difference.

GLUTEN-FREE FLOUR BLENDS

So, we know that the gluten-free flour or flour blend determines the rest of the ingredients (and their quantities) in a recipe and has a crucial impact on the success of a gluten-free bake. Throughout the book, I distinguish between a gluten-free flour and a gluten-free flour blend.

The former is only one ingredient – the gluten-free flour in question. That can be rice flour, almond flour, potato starch, arrowroot starch and many others (see Table 1, on page 17). The latter, on the other hand, consists of a mixture of gluten-free flours and sometimes also various additives, usually gums that lend gluten-free bakes their elasticity, as well as raising agents in the case of self-raising blends.

Nowadays, there are numerous gluten-free flour blends available in stores and online. To make this cook book truly comprehensive and complete, I have tested each recipe with a variety of gluten-free flour blends – both those you can buy ready mixed and those you can mix yourself (see Table 2, below).

There are several reasons why I've looked at both commercially available and DIY blends. The convenience of ready-mixed blends is a welcome help in hectic day-to-day life, and once you get accustomed to a specific blend, it becomes one of your best friends in the kitchen. Unfortunately, gluten-free flour blends too often get discontinued. And when that happens, it's like the loss of a dear friend, complete with soulful tears and abject dread of how you will live without them.

With a DIY blend, you need never rely on a brand or a store. Instead, you can always mix up a new batch whenever the need strikes. Plus, if you have intolerances to specific foods, you can modify the DIY blend to include or exclude certain ingredients, such as rice or maize flours or potato starch, which are common allergens.

Table 2 lists commercially available gluten-free flour blends that I've successfully tested for the recipes in this book, as well as two blends to mix yourself. To be as inclusive as possible, DIY blend 1 contains rice, potato and maize flour, the other avoids them. What's more, DIY blend 2 has a much lower starch content (only 40% compared to 80% in DIY blend 1) – this is to account for a common complaint with gluten-free baking: the starchiness of the flours and blends.

Table 2: Gluten-free flour blends and their ingredients.

BLEND	INGREDIENTS
Doves Farm Freee plain gluten-free flour	rice flour, potato starch, maize flour, buckwheat flour, tapioca starch
store-brand gluten-free flour blends (plain) [1]	rice flour, potato starch, maize flour
DIY blend 1	50% white rice flour, 30% potato starch, 20% maize flour [2]
DIY blend 2	40% tapioca starch, 30% buckwheat flour, 30% millet flour [2]

[1] At the time of writing, the following UK grocery stores sold their own branded gluten-free flour blends that I used in testing the recipes: Aldi, Lidl, Asda, Sainsbury's.

[2] For example, 100g of DIY blend 1 contains 50g white rice flour, 30g potato starch and 20g maize flour – the percentages always refer to grammes and never to volume measurements such as cups, as different flours have different densities.

There are further differences between the two DIY blends: while blend 1 is fairly neutral in flavour, blend 2 has an underlying nuttiness, which I love. In terms of colour, DIY blend 1 is pale yellowish (because of the maize) and DIY blend 2 is greyish-brown (because of the buckwheat and millet). The colours translate to the final bakes. While that's not really apparent in a chocolate cake, it is worth keeping in mind when making, say, a vanilla cake or white chocolate cupcakes.

Most recipes in the book refer simply to 'gluten-free flour blend' in their ingredients' lists (feel free to use any from Table 2), but you will also encounter recipes that reference specific gluten-free flours. In these cases, if you want to stray from these flours, make sure you follow the substitution advice below. Some recipes, particularly bread recipes, require a very specific flour profile with only little room for deviation if you want to end up with a delicious bake.

You will notice that Table 2 mentions only plain gluten-free flour blends – those that don't contain any raising agents mixed in. The reason is simple: using a plain blend and adding the raising agents yourself allows you more control over your bake, and enables you to fine-tune its rise and texture. Similarly, I recommend a gluten-free flour blend that doesn't contain xanthan gum. This way, you can tailor the specific amount of xanthan gum yourself for a perfect bake.

GLUTEN-FREE FLOUR SUBSTITUTIONS

You might wonder whether you can substitute one gluten-free flour within a blend for another. Yes – and no. The blends in the book should give the best bakes possible, so a substitution might alter the results. But, in general, all starch gluten-free flours can be swapped out for each other (with the exception of white rice flour, which is a bit of an outlier in the starch family). Things are trickier for protein flours.

I go into more detail about why that is on page 254, but for now, suffice to say that different protein flours impart different amounts of heaviness and flavour to bakes (especially bread) and are therefore divided into two groups, within which you can make swaps:

+ **brown rice and millet flours** (on the 'lighter' end of the heaviness spectrum and milder in flavour)
+ **buckwheat, maize, oat, quinoa, sorghum and white teff flours** (on the 'heavier' end and more intense in flavour)

Note that I have optimised the DIY gluten-free flour blends for the best results across a range of different bakes – so while I definitely encourage you to experiment with different flours (especially if you are intolerant to some of them or because they are unavailable to buy), do keep in mind that substituting flours can affect the texture, appearance and taste of the bakes. This is especially true for substituting protein flours because many of them are quite intense in flavour and colour.

OTHER PANTRY STAPLES

Knowing your gluten-free flours, flour blends, and binders is, of course, crucial for any successful gluten-free baking adventure. However, understanding and knowing which other ingredients to use is just as important – that's why I've included a list of my most important can't-live-without pantry staples below.

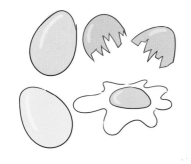

Eggs. In all my recipes, I use UK medium-sized eggs, which correspond to US large eggs. The average weight of whole, medium eggs is about 60g, or about 52g without the shell. The egg yolk weighs on average 16g and the egg white 36g. If you want (for whatever reason) to weigh out half an egg, do so by weighing out half the yolk (about 8g) and half the white (about 18g) separately – not by whisking the eggs and weighing out 26g. The latter is unlikely to give you the correct proportions of yolk and white. In most cases, your eggs should be at room temperature before you add them to batters and doughs. If you forget to take them out of the fridge beforehand (it happens to me all the time), just place them in a bowl of warm water for a few minutes.

Butter. Although I love salted butter spread generously on a slice of toasted bread (especially if the bread in question is the artisan dark crusty loaf on page 272), I always use unsalted butter in my baking. This is simply because it gives me better control over the bake. In most recipes, I'll use one of four forms of butter: softened; melted and cooled; chilled; or frozen and grated.

Softened butter refers to butter that's been standing out of the fridge for a while and when you press down on it with your finger, it should leave an indentation. It definitely shouldn't be hard to the touch, but it also shouldn't be verging on melted. If you forget to take the butter out of the fridge early enough, you can microwave it in 5–10 second bursts until you reach the correct consistency.

Melted and cooled butter is basically just butter that's been melted and cooled down until warm, so that it doesn't mess with any other ingredients it comes into contact with (especially eggs).

Chilled butter refers to butter straight out of the fridge. That means that even if you have to cube the butter for a particular recipe, you shouldn't leave it out at room temperature afterwards. Instead, keep it in the fridge until just before you need to use it.

Grated frozen butter is butter that's been frozen for a minimum of 2–3 hours, or preferably overnight, and is then coarsely grated. This leaves the butter in a paradoxical state of being both extremely cold and very malleable, which is perfect for quickly laminating doughs, such as rough puff pastry (see page 203).

Oil. Although butter is definitely my fat of choice in baking, sometimes using oil is best. If I specify a neutral-tasting oil, I usually mean sunflower or rapeseed oil, but vegetable oil works just as well. If you want to introduce a flavour along with the oil, olive oil is usually the way to go, but feel free to use other flavoured oils that you like.

Sugar. You'll notice I prefer fine over coarse sugar in my recipes – for example, I'll use caster sugar over granulated sugar in most cases. Caster sugar comprises smaller crystals that are easier and so quicker to dissolve than the larger particles of granulated sugar. In bakes, such as chocolate sponges or French crêpes, that have a lot of liquid, this isn't hugely important, but in bakes like cookies or pastry, caster sugar gives a more even, homogeneous crumb. That said, when it comes to sprinkling sugar on top of muffins or egg-washed pastry, granulated sugar is the way to go – for pretty much the exact opposite reasons. Here, we don't want the sugar to dissolve but to form a sparkling, slightly crunchy crust. In addition to white sugar, I use (soft) light and dark brown sugars, which add both flavour and moisture. The extra moisture is especially important in bakes where you want a fudgy or gooey texture, such as in brownies or chocolate chip cookies.

Honey + maple syrup. Although white and brown sugar are my go-to sweeteners in the majority of cases, honey, maple syrup and other liquid sweeteners can be fun to play around with. In fact, you can often use them in place of regular sugar, with a few important exceptions:

+ where creaming butter and sugar, or whisking together eggs and sugar is the central reason why a recipe works (like the shiny top brownies on page 126),
+ where sugar controls the way a bake behaves in the oven (for example, sugar is crucial for chocolate chip cookies spreading out during baking; see page 152),
+ where you don't want to introduce extra moisture into the recipe (notably in bread and the other dough recipes on pages 266–317).

Salt. I don't think there's a single recipe in this book that doesn't contain at least a pinch of salt (with 'salt' referring to sodium chloride, NaCl, and a few other molecules and compounds mixed in depending on the type of salt). That's because salt is a natural flavour enhancer, even in very tiny amounts – and it's especially important when dealing with chocolate or peanut butter, where that one pinch takes the bake to a whole new level. Similarly, adding too little salt to a loaf of bread will make it taste rather bland and cardboard-like (no matter how wonderful its texture is). I use sea salt in my kitchen, both in the fine, free-flowing form and in the flaky form, the latter especially for sprinkling on top of cookies and brownies – although most other types of salt will work just as well in the recipes.

Raising (chemical leavening) agents. When it comes to using baking powder and bicarbonate of soda, the most important thing to understand is their composition.

Bicarbonate of soda is a pure compound – sodium bicarbonate, $NaHCO_3$ – a base or alkaline substance that reacts with acids to release a gas (carbon dioxide; CO_2), with water and a salt forming as the side products (where 'salt' isn't necessarily table salt or sodium chloride). Therefore, the presence of an acidic ingredient is crucial for bicarbonate of soda to be active and to make the bakes light and fluffy. This acidic ingredient can be any one of a variety of things, such as lemon juice, yoghurt, soured cream or even chocolate (whose pH falls on the acidic end of the spectrum). Because it's a pure compound, bicarbonate of soda doesn't really lose its activity and, unless otherwise contaminated, has a pretty much indefinite shelf life.

Baking powder, on the other hand, is not a chemically pure substance. Instead, it's composed of bicarbonate of soda, an acidic component and anti-caking/filler agents such as rice flour or cornflour to prevent it from clumping together and to keep the bicarbonate of soda and the acidic component as dry as possible (by preferentially absorbing any moisture present) – we'll get to why that's important in a minute. The acidic component can be cream of tartar, a phosphate or a sulphate. For the base (bicarbonate of soda) and the acid to react and release carbon dioxide, they need to mix together in an aqueous medium and preferably in the presence of heat (as the heat speeds up the reaction and also vaporises the water produced, so increasing the number of 'gas bubbles' in the bake to result in a light and fluffy texture). That's why keeping moisture away is so important and why we add filler agents. A humid and warm or hot environment can significantly shorten the shelf life of baking powder. To test whether baking powder is still active, just mix it with some water: if it gets frothy and bubbles up, it's active and okay to use.

Yeast. While bakes such as cakes and cupcakes rely on chemical leavening agents for their rise and open crumb, bread looks to yeast as its leavening agent. Now, I know what you must be thinking: using only yeast in gluten-free bread? Surely, we need some baking powder to help it along. Spoiler alert: not at all – but more on that in the bread chapter. For now, let's talk about the type of yeast my recipes use: active dried yeast. Although the way active dried yeast is processed means that it's slightly less active than instant yeast (so you need slightly more per 100g of flour), I love that you need to activate it in some warm water or milk and so can actually see it getting all nice and frothy. In addition to confirming your faith in its ability to do its job (and that it's not past its sell-by date or inactive for some other random reason), activating the yeast can become almost a part of the ritual that is baking the weekly loaf of gluten-free bread. That said, you can use instant (rapid- or quick-rise) yeast in most of my recipes, just reduce the weight to 75% of that listed, so that you use 0.75g instant yeast for every 1g active dried yeast. Unlike active dried yeast, instant yeast doesn't need to be hydrated and you can add it directly to the dry ingredients. (If you want to use fresh yeast, you'll need double the weight of the active dried yeast listed in the recipe.)

Chocolate. It goes without saying: use a high-quality chocolate for baking, one you would genuinely enjoy eating. Most of my recipes use dark chocolate with cocoa-solid percentages in the 60–70% range, but there are quite a few that use white chocolate and some with milk chocolate (the hazelnut and milk chocolate cake on page 54 comes to mind… and, oh gosh, it's a winner). In recent years, the quality of chocolate accessible to the average baker has increased drastically, so there's really no excuse for using anything that's subpar.

The reason I specify the cocoa percentages is two-fold. First, when it comes to things like chocolate ganache or glaze, the cocoa content drastically affects consistency. A chocolate glaze that is perfectly pourable with a 60% chocolate can be quite noticeably thicker with a 70%+ chocolate (depending, of course, on the other ingredients). Second, there's the flavour or, more specifically, the sweetness of the dessert. For things like the chocolate-mousse filling in the hot chocolate tart (see page 244), I like to use a chocolate with a lower percentage of cocoa solids, so that there's no need to add additional sugar (as might be necessary with a darker, more bitter chocolate).

Dutch processed cocoa powder. In all the recipes that require cocoa powder, I use Dutch processed cocoa powder, not to be confused with natural, unprocessed cocoa powder. Natural cocoa powder is acidic, with a pH of 5–6 (remember, neutral pH is 7), which gives it a sharper, fruitier taste. Its acidity also means that it will react with baking powder and bicarbonate of soda and work to additionally aerate the bake. Dutch processed (also known as 'alkalised') cocoa powder goes through an alkalising process that neutralises its acidity to pH 7. At the same time, the process gives it a deeper, darker colour and a far richer flavour we commonly associate with chocolate. In your general baking, I wouldn't advise trying to substitute one for the other if the recipe specifies which one to use – messing with the acidity levels of a bake will inevitably change how the chemical leavening agents react both in and out of the oven and, considering the recipe has been optimised for a certain composition, that change probably won't be a positive one.

Vanilla bean paste. When it comes to adding a gentle vanilla undertone (or, if needed, a powerful vanilla punch as for the simple vanilla cake on page 50 and the vanilla cupcakes on page 92), I will always reach for vanilla bean paste over vanilla extract. In addition to giving desserts that wonderful speckled appearance from actual vanilla beans (which you don't get in an extract), I prefer the intensity and the quality of flavour it provides. I also find that you need to use far less of it than you would of the extract – so while it might be a bit of an investment compared to the extract, I think it's one that rewards you several times over.

KITCHEN TOOLS

With things that go into the bakes out of the way, let's have a look at the indispensable tools that make gluten-free-baking perfection possible. These are items I use in my kitchen on a daily basis.

CASTER SUGAR (OR MAYBE SALT)

PEANUT BUTTER ... PROBABLY

Digital scales. This holds the #1 spot in the kitchen tools' list for a very good reason. With the exception of adding things like baking powder, bicarbonate of soda, xanthan gum and salt by teaspoons (and even that disappears come the bread chapter, where grammes are king), I weigh pretty much everything else. Yes, that includes liquids like water, milk and oil. When it comes to precision, nothing – and I mean nothing – matches simple digital scales. Weighing out 137g water is a piece of cake, but trying to measure out 137ml in a jug is a whole other story. (Unless you have a lab-grade measuring cylinder, in which case good on you.) So, I really can't emphasise this enough: ditch unreliable volume measurements and get some digital food scales. Bonus points if the display shows decimal places.

OVEN THERMOMETER

Oven thermometer. Let me tell you a story: during my university years, I lived in student accommodation and, one day, tried to bake a cake in the oven there. After over an hour, it came out raw. Why? The oven was set to 180°C, but the actual oven temperature was a measly 130°C. That's right, it was 50°C out of calibration. The moral of the story is that you really shouldn't blindly trust the settings on your oven – it's much better to check how well it's calibrated using an oven thermometer, and then either re-calibrate if possible, or simply remember which temperature settings correspond to which actual temperatures. My current oven is 10°C out of sync, so I know I need to set it to 190°C when I want to bake something at 180°C. A final note on oven temperatures: don't just measure the temperature the oven heats up to initially. Instead, pay attention to how the temperature varies with time and especially after you've put something in there (even an empty baking tray). This temperature is far more important than the temperature the oven reaches and maintains for the initial 5 minutes or so.

Digital food thermometer. In addition to being crucial when making caramel or tempering chocolate, a food thermometer will serve you well for everything from measuring the internal temperature of bread or pies, to monitoring the temperature of egg whites as you cook them with sugar for Swiss meringue buttercream. That's why I prefer a digital thermometer with a metal probe as opposed to an old-fashioned analogue jam thermometer, the shape of which limits its usefulness.

Mixing bowls. I possess a rather unreasonable number of bowls, and each time I think there's one too many, I still run out of clean ones when I go on a crazy baking spree. When it comes to bowls, consider two things: their size and the material they're made from. I have a variety of different sizes, from small ones for cracking individual eggs to huge ones for when I want to make 2 kilos of cookie dough (we all have those times, right?). The material of a bowl should be heatproof, sturdy and durable – I swear by Pyrex glass and stainless steel.

Stand mixer. This one goes without saying. Although a hand mixer works in a pinch, a stand mixer is useful for a wider range of recipes (from making meringues and buttercreams to kneading dough and preparing shortcrust pastry). Plus, the many speed settings allow you to control the aeration of creams and batters – which is crucial in many recipes, from making cheesecake filling to whipping cream.

Food processor. I use this important bit of kit for everything from making the base for lemon and s'mores bars (see pages 188 and 184) to grinding hazelnuts and making frangipane.

Whisks, wooden spoons + spatulas. These are no-brainers. You need them for everything. At this very moment, I have more than 15 spatulas in my kitchen, and I still end up running out of clean ones (don't ask me how). The recipes guide on which implement to use when but, in general, use a whisk to aerate the mixture or to break up clumps in a batter quickly and efficiently. When you don't want to incorporate too much air, a wooden spoon is the way to go. And nothing beats a rubber spatula when it comes to folding dry ingredients into a fluffy mixture of eggs and sugar (and chocolate, in the case of brownies) without deflating it too much. Or, for that matter, when it comes to scraping every last bit of whatever deliciousness you're baking into the tin.

Baking sheets, trays + tins. Here's an overview of the main baking containers I use in this book:

+ 20cm anodised aluminium round cake tins (5cm deep)
+ 20cm anodised aluminium square baking tin (5cm deep)
+ 23cm pie dishes (preferably metal, 3cm and 4cm deep)
+ 23cm loose-bottomed tart tin with a fluted edge (about 3.5cm deep)
+ 20cm round springform tin (7cm deep)
+ 900g loaf tin (23 x 13 x 7.5cm)
+ 1 shallow baking tray (25 x 35cm, about 2.5cm deep)
+ 1 deep baking tray (23 x 33 x 5cm)
+ baking sheets (at least 2, with and without a lip)
+ 12-hole muffin tin

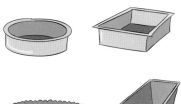

I've developed the recipes using the specified tins and dishes – using a different tin size, material and/or colour can produce slightly different results and will affect the baking times. That's why I recommend using physical indicators (see page 30) to judge whether a bake is done.

Baking paper. I know many prefer silicone baking mats, but I have always used baking paper, which you can cut to shape and size. If silicone mats are your best baking friend, though, go ahead and use those – just note that some recipes might give slightly different results (for example, cookies usually spread out more on a mat). Beyond just using it to line baking sheets and tins, baking paper comes in handy as a rolling surface for pie crust, other pastry, and gluten-free enriched dough, especially because it allows you to slide everything on to a baking sheet and pop it into the fridge if it needs a quick chill.

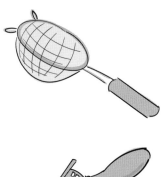

Fine-mesh sieve. From sifting dry ingredients to remove any stray clumps and introduce some extra air into the mixture, to passing cooked raspberries through to strain the juices on their way to becoming a delicious reduction, fine-mesh sieves play a big part in baking my recipes. I recommend having a few sizes on hand – in terms of both the sieve diameter and the holes in the sieve mesh.

Ice-cream scoop(s). Regardless of whether you call them ice-cream scoops or cookie scoops, these are incredibly useful for everything from scooping cookie dough to quickly portioning out cupcake batter into cupcake liners. The scoop sizes are usually described by the tablespoon volume measurement (that is, 2-tablespoon scoop, 3-tablespoon scoop and so on). I've specified where relevant.

Pastry brush. While you may mainly use a pastry brush for glazing – and this could be egg washing a pastry or brushing an enriched dough with butter before baking, it's also useful for gently brushing off excess flour from your pastry or dough. I vastly prefer brushes made with natural bristles over silicone ones, but feel free to use whichever feels more comfortable to you – and whichever you have to hand.

Offset spatulas. Ah, the things you can do with an offset spatula. First, the obvious ones: swirling frosting, icing or ganache on top of cakes, and spreading filling on top of cinnamon rolls or chocolate babka. But a large offset spatula is also very handy for sliding beneath rolled-out cookie dough before you cut out the cookies, so you're 100% sure none of them will stick. Then, you can use it to transfer unbaked cookies from the rolling surface to the baking sheet, and (a few minutes later) baked cookies from the baking sheet to a wire (cooling) rack – and myriad other little things. Because offset spatulas are so versatile, I recommend having at least two sizes in your kitchen: a large one for general use and a small one for more detailed and finicky work.

Piping bags + nozzles. You will mainly have to pipe three things in this book: buttercream (on to cakes, cupcakes, cookies and the occasional brownie), choux batter (to make choux puffs and éclairs) and filling (into choux bakes and doughnuts and on to millefeuille and cookies). When it comes to frostings and fillings, mine are just suggestions as to the piping nozzles you should use – you don't have to follow them slavishly. However, for piping choux batter, using the correct piping nozzle is very important, as it dictates the way the choux buns and éclairs expand in the oven, which in turns controls their shape and whether or not they crack. Here's an overview of the piping nozzles called for in the recipes:

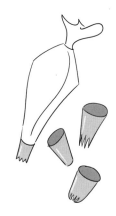

+ **medium open star nozzle** for piping frosting on to cupcakes
+ **large open star nozzle** for piping frostings and fillings, and éclairs
+ **French star nozzle** for piping éclairs
+ **rose petal nozzle** for filling éclairs
+ **large round nozzle** for piping frostings and fillings, and choux buns
+ **small round nozzle** for delicate decorating work and for filling baked doughnuts

Wire (cooling) racks. When cooling your bakes to room temperature, it is crucial to ensure proper air circulation around them, as this both speeds up the cooling process and prevents any condensation build-up, which could make bakes soggy or otherwise spoil their texture. Wire cooling racks are perfect for cooling – but they're equally great when it comes to glazing cakes, as they allow any excess glaze to drip through and away rather than accumulate around the cake in a puddle.

Rotating cake stand. My cake-decorating experience went from 'meh' to 'this is the best thing ever' after I got my hands on a rotating cake stand (also known as a cake-decorating turntable). If you like to make cakes, and especially if you make a lot of them, a rotating cake stand is a worthwhile investment.

Rolling pin. I could go on and on about what makes the perfect rolling pin, but, in the end, the big question is how it feels in your hands and how much control it gives you over how thinly and evenly you can roll out pastry and dough. There are tens of different types of rolling pin out there – with or without handles, straight, tapered, wooden, ceramic – and you're bound to find one that makes rolling a breeze for you. Just make sure it's long enough to handle larger amounts of pastry and dough (it should be at least 24cm long).

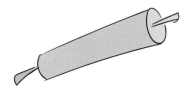

Foil. This has one main purpose in my kitchen: to control the rate at which bakes are browning. Placing a sheet of foil (shiny side up) over a bake blocks out much of the radiative heat to slow down the rate of browning. This allows you to tune the colour of your bake irrespective of how much longer it needs in the oven to reach perfection.

Cling film. I use cling film for everything from wrapping up flaky pie crust and rough puff pastry to prevent them from drying out in the fridge to using it as the rolling surface for pastry (it's incredibly helpful for transferring pastry into tart tins and baking dishes with minimal chance of tearing). And if you can get your hands on heat-resistant cling film, you can even use it instead of baking paper in blind baking.

Toothpicks + skewers. Two words: toothpick test. Read all about it on page 30.

Bench scraper. Decorating cakes, shaping bread, dividing dough into even portions… there are endless uses for a bench scraper. But probably the most important one in my kitchen is when it comes to cleaning time: a bench scraper is the perfect thing to get any stubborn dough or batter remnants off your kitchen work surface. When it comes to materials, a metal scraper is sturdy and durable, while a flexible plastic one is amazing at getting slightly sticky dough out of bowls and other bulk-proving containers.

Proving baskets. You will need only one size of proving (bread) basket if you make the artisan crusty loaf recipes (on pages 266 and 272) from this book, without scaling the recipes up or down – an 18cm proving basket that's 9cm deep. I tend to use it without the cloth cover that usually comes with bread baskets, mostly because I love the circular pattern left by the basket itself. Before proving the bread, it's important to dust the basket generously with flour (I mostly use millet flour for this) to prevent the bread sticking to it.

In terms of upkeep: do not, under any circumstances, ever, wash your proving basket. Just turn it upside down after you've finished using it and gently tap it on the counter to shake off any clumps of flour. Eventually, after repeated use, it will become almost 'seasoned' with a layer of flour – that's perfectly okay and nothing to worry about. Furthermore, do not keep it in an airtight container. The basket absorbs a certain amount of moisture from the bread, and keeping it in a closed environment is recipe for a mouldy disaster. Just keep it in a drawer or on a shelf, lightly covered with a tea towel to prevent it from getting dusty.

Cast-iron Dutch oven, combo cooker or skillet. These are indispensable when it comes to baking boules (round loaves of crusty bread) and I have included instructions for using all of them in my recipes. Cast iron is excellent at retaining and radiating heat, which gives bread the most beautiful crust and promotes oven spring (see pages 260–261). Beyond bread, I use a cast-iron skillet to make skillet cookies, and you can even use it to bake pizza.

Lames + other scoring razors. Scoring (see page 260) is incredibly important for controlling the way your bread expands in the oven – plus, it looks pretty. You can use a variety of scoring tools, but for a beginner the combination of a simple lame (double-sided razor blade intended for scoring bread) and holder is an excellent starting point.

THE TOOTHPICK TEST + FAN OVENS

We've touched on the problem of incorrectly calibrated ovens, which can drastically affect the baking times in the recipes. What's more, different oven volumes (what we commonly think of as our oven size) change the proximity of a bake to the heating elements, which in turn affects the rate at which things bake and brown. Then, there's also the question of whether or not to use a fan oven, which again changes the rate at which things cook through and dry out (see opposite).

So, the baking times in the recipes are very good rough guides, but it's far more reliable to use physical indicators for whether or not your bakes are done. These indicators vary according to the type of bake, but here's a list of the most important ones:

1. **Toothpick test.** Here, you insert a toothpick or skewer into the centre of your bake for about 2 seconds and then see how it comes out. Completely clean, with a few stray crumbs attached, or covered with half-baked batter? The state of that humble toothpick tells you more than you can imagine about the doneness of your bake. Whenever you're baking cakes, muffins or brownies, this is the one trick you need to have up your sleeve. Where relevant, the recipes will tell you what to look for.

2. **Weight loss.** This is especially important when baking gluten-free bread, where other tests might mistakenly tell you your loaf is done. By weighing the loaf, you can calculate the percentage weight loss and accurately determine whether it needs a bit longer in the oven. Where relevant, the recipes will tell you what sort of weight loss to aim for.

3. **Colour.** When it comes to the colour of your bakes, there are two sides to the story. When dealing with cakes, cupcakes, muffins and bread, colour isn't a very reliable indicator of whether or not a bake is done. Depending on the amount of sugar, baking time required, depth of baking tin and myriad other factors, the top of your bake might start browning well before it's actually done. (In which case: foil – shiny side up – to the rescue.) In the case of (non-chocolatey) cookies, however, colour is an excellent indicator. For bakes such as sugar cookies (see page 158), shortbread biscuits (page 161), peanut butter sandwich cookies (page 169) and rosemary crackers (page 174), colour is the easiest way to test for doneness (provided you bake them at the correct temperature). I've specified the colour you're looking for in the final bake in each recipe, as relevant.

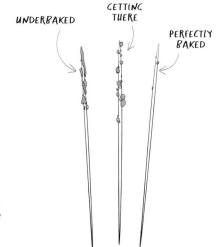

UNDERBAKED
GETTING THERE
PERFECTLY BAKED

Fan ovens in gluten-free baking

In general, unless it's the only option available, I don't recommend using a fan oven for gluten-free baking, which is why you'll find only one oven temperature in my recipes. Fan ovens tend to have the unfortunate effect of drying out baked goods. This may be a small, almost insignificant effect in wheat-based baking, but in gluten-free baking you are constantly battling the notion of a dry, crumbly bake.

A fan oven (also known as a fan-forced oven) uses a fan to circulate the hot air generated by the heating element. Compared to a conventional oven, a fan oven typically creates more heat as well as a 'breeze' in the oven. While more heat means faster baking, the combination of more heat and a breeze results in more moisture evaporation and therefore a drier bake. This is especially noticeable in bakes that require a longer baking time, such as deep loaf cakes or decadent chocolate cakes. And the thought of putting so much effort into making a chocolate cake moist and luscious, only for it to be dried out by the fan setting – well, it's just heartbreaking. Another disadvantage of the high heat that fan ovens generate is accelerated browning, which can mislead you into thinking your bake is done, when actually it may need a further 10–15 minutes.

Therefore, I like to stick with the conventional oven – the only time I use the fan setting in my kitchen is to crisp up roasted potato wedges. That's not to say gluten-free baking won't work in a fan oven. Much like a gluten-free flour blend you know and trust, your oven is (or at least should be) your close friend. You know its quirks and whimsies – and if the fan setting is its strong point and you know the ins and outs of it, go ahead and turn that fan on. In that case, as a good rule of thumb, reduce the conventional oven temperatures in the book by 20°C when using the fan oven. (Conversion tables are given on pages 370–374.)

COOK'S NOTES

Before you delve into the recipes, there are a few details you need to know. These apply to all the recipes in the book:

+ Use any of the gluten-free flour blends listed in Table 2 (see page 19), unless otherwise specified.
+ In cake, cupcake and muffin recipes, you can replace almond flour with an equal weight of gluten-free flour blend (but, see page 16).
+ Baking powder is typically gluten-free, but always check the label.
+ Use medium eggs, fresh herbs and medium-sized fresh fruit and vegetables, unless otherwise specified.
+ If the temperature of typically refrigerated ingredients isn't specified, use cold or at room temperature.
+ Spoon measures are level, not heaped or rounded.
+ The oven temperatures are for a conventional oven, not fan.

CAKES

Baking a good cake is much like having a real-life superpower.
A beautiful cake will stop the world, for a short moment, in a state
of happy, blissful surprise. It doesn't matter whether it's a simple
traybake or a three-tiered, elaborately decorated affair: show up with
a cake and people will stop. And look. And smile. Turning invisible,
reading minds or being able to fly? That's for amateurs.

Gluten-free cakes have that same magical stop-stare-smile effect.
Contrary to popular belief, being gluten-free doesn't prevent cakes
from being moist, delicate, airy, soft, luscious and all the other
things you might want a cake to be. In fact, achieving the perfect
gluten-free crumb is often easier than in wheat-based baking,
where the presence of gluten can result in a dense or gummy
texture if you overmix the batter. To get the most delicious results,
however, and to do it in a consistent and reproducible manner,
there are a few details you need to keep in mind.

THE TYPES OF CAKE YOU'LL FIND IN THIS CHAPTER

I've focused this chapter on what I believe to be the fundamental cake recipes – those that, once you know them, allow you to take on any cake-making challenge, no matter how complicated. In addition to the basic chocolate and vanilla cake recipes (refined to the very pinnacle of deliciousness), there are recipes that look at the effects of adding an acidic element, introducing extra moisture into the batter, and using different tin sizes and shapes, and that take on the scary prospect of rolling the sponges into a (crack-free!) Swiss roll or roulade.

WHY MAKING GLUTEN-FREE CAKES IS RATHER EASY

Cakes are forgiving. What this means is that most store-bought gluten-free flour blends will give you a good rise, a moist crumb and an overall delicious cake – if you, of course, follow all the other nuggets of science and wisdom laid out in this section. The reason for this is fairly simple and it all comes down to maths. And rather easy maths at that. What I'm referring to is the percentage of flour among all the other ingredients that go into a bake – as laid out in Table 3.

Table 3: Percentages of flour in typical cake, cupcake, muffin, brownie, cookie, pie crust, shortcrust pastry and bread recipes. (The values are estimates based on representative recipes from each category.)

TYPE OF BAKE	PERCENTAGE OF FLOUR
brownies	9%
cake	27%
cupcakes	27%
muffins	33%
bread	42%
cookies	43%
pie crust	47%
shortcrust pastry	54%

Apart from brownies, cakes and cupcakes contain the smallest percentage of flour of all the bakes you will find in this book. It's therefore hardly surprising that the exact gluten-free flours or flour blends don't play as large a role as they do in, for example, cookies or pie crust.

I have thoroughly tested all the cake recipes in this chapter with all the gluten-free flour blends outlined in Table 2 on page 19 – with a 100% success (and deliciousness) rate.

HOW TO MAKE (AND KEEP) YOUR CAKES MOIST

Gluten-free flours and flour blends tend to absorb more moisture both before and during baking than wheat flour, which can potentially result in a dry and crumbly bake. To avoid this, it's important to increase the wet:dry ingredients' ratio. Wet ingredients include eggs and any liquids (milk, water, yoghurt, brewed coffee, fruit purées and juices), whereas dry ingredients include the gluten-free flour or flour blend, and the cocoa powder, ground nuts or similar. If you're used to wheat-based baking, the amount of liquid, in particular, might on occasion feel almost counterintuitive. But, fear not – it's this additional liquid that helps create a wonderfully moist, melt-in-the-mouth cake crumb.

Increasing the wet:dry ratio deals primarily with the issue of how moist the cakes are immediately after baking and for a few hours afterwards. When it comes to the rate at which the sponges dry out (even when in a closed container, wrapped in cling film or protected in a layer of frosting), we need a different trick up our sleeve.

Despite their high moisture-absorption capacity, gluten-free flours are also surprisingly (and unfortunately) rather good at losing that moisture, resulting in cakes that become noticeably dry and crumbly after a day or two. To slow down this rate of drying out, I like to substitute a tablespoon or two of the gluten-free flour blend for almond flour, which works wonders at keeping the cakes moist and fluffy for days, mostly because of its higher fat content as compared to gluten-free flours. This gives the sponges a moister and richer mouthfeel without actually making them dense, heavy or oily (as, for example, increasing the amount of butter might do). If you have a nut allergy or similar, and therefore can't use almond flour, you can substitute it with an equal weight of the gluten-free flour blend.

Xanthan gum also deserves a mention here, as its purpose to mimic gluten isn't restricted to acting only as a glue that makes the sponges significantly less crumbly. It also absorbs and retains moisture to help them stay nice and moist – in tests, xanthan gum has been able to absorb more than 20 times its weight in water!

Let's finish with an almost stupid-simple, yet often overlooked solution to getting moist cakes: reducing the baking time. It's possible to overbake a sponge in a matter of minutes, so keep an eye on your cakes a few minutes before you hit the suggested baking time. Depending on how your oven is calibrated (see page 25), a cake may take a few minutes more or fewer to be properly baked.

And don't forget the toothpick test (see page 30)! For cakes, always look for a mostly clean toothpick with maybe a stray crumb or two attached.

RAISING AGENTS GO INTO THE DRY INGREDIENTS

So, you've chosen the correct ingredients in the correct relative proportions – great. But you're nowhere near done yet. How you combine these ingredients is just as, if not more important. Before we get down to the nitty-gritty of mixing and whisking the batter, let's clear one thing up: raising agents (that is, baking powder and bicarbonate of soda) always go with the dry ingredients. For the love of delicious fluffy cake, don't go mixing them into the wet ingredients!

I've seen people mix baking powder into eggs and milk, thinking it will give them a 'more homogeneous dispersion' of raising agents in the final batter. (It doesn't.) This thinking is similar to the idea of 'activating baking powder and bicarbonate of soda in lemon juice or vinegar before adding it to the batter' – please don't do that. Putting either raising agent into an acidic medium results in immediate gas release and salt formation. This destroys the purpose of the raising agent, activating it long before you want it working. What you want is a delayed reaction – in the oven, to give a tender, fluffy crumb.

Mixing raising agents into the wet ingredients poses a similar problem, especially in the case of baking powder. If we look at its composition, baking powder is a mixture of bicarbonate of soda with an acidic component such as cream of tartar (baking powder therefore doesn't need a separate source of acid in the cake batter in the way bicarbonate of soda does). When dissolved in a liquid such as water and in the presence of heat, the two components react to release gas and make our cakes fluffy. However, the reaction can occur even in the absence of heat – try mixing a bit of baking powder with a few drops of water. The mixture will bubble up, releasing gases.

Adding baking powder to wet ingredients, therefore, does exactly the same thing, making the bicarbonate of soda in it react with the acidic component – and, bye-bye, there goes the leavening activity. If you're

Baking powder is a mixture of bicarbonate of soda and an acidic component. When it is mixed with water (or other liquids) and subjected to heat, the reaction results in immediate gas release and salt formation.

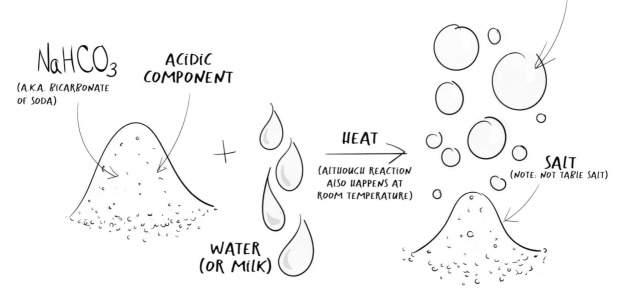

$NaHCO_3$
(A.K.A. BICARBONATE OF SODA)

ACIDIC COMPONENT

+

WATER (OR MILK)

HEAT
(ALTHOUGH REACTION ALSO HAPPENS AT ROOM TEMPERATURE)

CARBON DIOXIDE & WATER VAPOUR

SALT
(NOTE: NOT TABLE SALT)

using only pure bicarbonate of soda, you might get away with it, assuming you're not adding any sources of acid to the wet ingredients. But considering that even milk is slightly acidic, it's better to stick to adding all chemical leavening agents to the dry ingredients.

Of course, when you mix the dry ingredients with wet, the raising agents will still come into contact with liquids and acids. But considering their relatively low concentration within the dry ingredients, the probability of that contact, and therefore of them losing their activity before it's needed, is significantly reduced (and the amounts of baking powder and/or bicarbonate of soda listed in the recipes account for small losses anyway). Nonetheless, this is why it's best to bake the cake batter straight away, rather than letting it sit around in a bowl or baking tin. (If you have only one baking tin and the recipe requires two, it's best to prepare only half the batter, bake it, and then make and bake the other half.)

THE THREE MIXING METHODS

With cakes based on using softened butter (rather than melted butter, or oil) there are three mixing methods to choose from:

+ the standard creaming method,
+ the reverse creaming method,
+ the all-in-one-method.

Let's deal with the last one first. As its name suggests, the **all-in-one method** involves placing all the sponge ingredients in a bowl and whisking until you get a smooth batter. An obvious advantage of this method is its speed and simplicity, as well as the fact that there's only one bowl to wash up (always a welcome surprise). However, it doesn't offer much in terms of controlling the way the ingredients come into contact with each other and interact, and it is the method most likely to give you stray patches of flour or butter. In general, I reserve it for quick, everyday bakes, such as the Victoria sponge cake on page 328.

The **standard creaming method** is the one bakers are usually most familiar with. You cream the butter with sugar into pale, fluffy submission, then slowly incorporate the eggs to create a voluminous emulsion. Finally, you add the dry and any remaining wet ingredients (such as milk), usually in alternating batches. As well as being incredibly satisfying to observe (who doesn't love a soft, airy cake batter?), this method produces very light and fluffy cakes with a delicate crumb. It can, though, cause an uneven aeration (a cake interior filled with holes of different sizes), a tendency towards a domed top, a greater likelihood of the cake collapsing if you open the oven door during baking, and (specifically for gluten-free cakes) a slightly greater rate of drying out (mostly because of a more open crumb).

STANDARD CREAMING METHOD

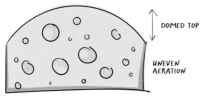

DOMED TOP

UNEVEN AERATION

REVERSE CREAMING METHOD

↕ FLAT TOP

EVEN AERATION

A look at how the two creaming methods affect the texture and appearance of baked sponges, and what the cake batters look like at the microscopic level – both before and during baking. The reverse creaming method is far better for achieving the coveted flat top and even aeration of the perfect sponge.

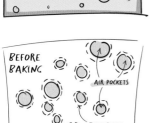

BEFORE BAKING

AIR POCKETS

BUTTER FAT PARTICLES

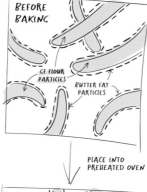

BEFORE BAKING

GF FLOUR PARTICLES

BUTTER FAT PARTICLES

PLACE INTO PREHEATED OVEN

PLACE INTO PREHEATED OVEN

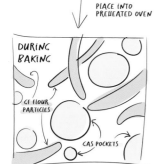

DURING BAKING

GF FLOUR PARTICLES

GAS POCKETS

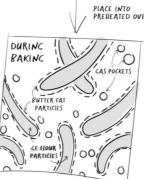

DURING BAKING

GAS POCKETS

BUTTER FAT PARTICLES

GF FLOUR PARTICLES

The main reason for these potential problems lies in the large amount of air, in the form of tiny air pockets, incorporated into the batter during the creaming method. During baking, these air pockets expand, creating holes in the crumb and a domed top. Of course, we want gas bubbles to be present or to form in the cake as it bakes – they are the source of that beloved fluffiness and tenderness. But, we also want to control the amount, size and distribution of the bubbles, which is almost impossible with the standard creaming method. Mostly, I use the standard creaming method only for very moist, buttery cakes on the denser side of the fluffiness scale, such as bundt cakes (see page 62) or lamingtons (see page 369), where the larger amounts of butter counteract most issues of uneven aeration and likelihood of collapse.

Compared to the standard method, the **reverse creaming method** (pioneered by Rose Levy Beranbaum) gives a finer cake crumb with a velvety texture that melts in the mouth – and no doming or uneven aeration in sight. Here, you work the butter into the dry ingredients until it resembles breadcrumbs or coarse sand. Then, you add the wet ingredients to form a smooth batter. This leads to the (gluten-free) flour being coated in butter particles with little air getting trapped in the mixture. The rise in a sponge prepared by the reverse creaming method therefore comes almost exclusively from the raising agents (and to a smaller degree from the eggs), rather than from the air bubbles you incorporate as you prepare the batter.

This is the method I use for the majority of my cakes – the greater control it gives over the cake structure and appearance outweighs any concerns about reduced fluffiness. And in any case, it is easy to tweak the lightness of a cake using other (more precise) means, such as adding extra egg whites, slightly increasing the raising agents or adding acidic ingredients, such as lemon juice, buttermilk or yoghurt.

THE (VERY IMPORTANT) ROLE OF THE CAKE TIN

I really cannot overemphasise the importance of choosing the right tin for baking your cakes – or, perhaps somewhat more importantly, knowing your way around the tins you have. The size, depth and shape, and even the material, of the cake tin all play a large part in how your cakes bake: they affect the baking time, the way a cake rises, the rate at which the edges caramelise, and myriad other little things.

While this isn't hugely important if you follow the recipes to the letter, as they're all optimised for the tin sizes I've given you, I expect you might not have the same cake tins in your kitchen as I do in mine. Or, you might want to bake a 20cm cake instead of a 23cm one, or a square cake instead of a round one.

In general, you should be able to switch between different cake tins effortlessly – as long as you're aware that you will have to adjust the baking time accordingly. That's because the rate of heat penetration into the cake batter, and therefore the rate at which the cake bakes, varies hugely with the parameters of the cake tin. Here are the most important points to keep in mind:

+ **Increasing the diameter of the cake tin while keeping the amount of batter constant** will create a thinner layer of cake batter in the cake tin, reducing the baking time – and vice versa for a cake tin with a smaller diameter.

+ **Increasing the diameter of the cake tin while also proportionately increasing the amount of batter** to create a layer of batter of the same thickness will increase the baking time.

+ **Loaf tins, bundt tins and other deep cake tins** will always require a longer baking time because of the thicker layer of cake batter.

+ As a rule of thumb, **the surface area of a square tin is approximately 25% larger than the surface area of a round tin**

HEAT PENETRATION

INCREASE BAKING TIN DIAMETER →

LEAVE AMOUNT OF BATTER THE SAME

SHORTER BAKING TIME

Changing the size and shape of the baking tin from the one listed in the recipe to another potentially changes the thickness of the sponge batter in the tin. This, in turn, affects the baking time and/or the amount of batter you need to achieve perfect results.

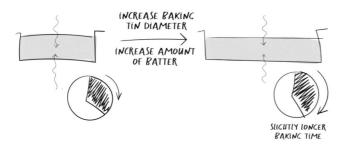

INCREASE BAKING TIN DIAMETER →

INCREASE AMOUNT OF BATTER

SLIGHTLY LONGER BAKING TIME

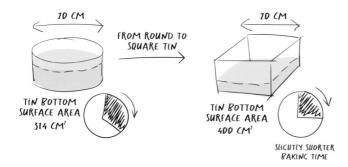

20 CM

FROM ROUND TO SQUARE TIN →

20 CM

TIN BOTTOM SURFACE AREA 314 CM²

TIN BOTTOM SURFACE AREA 400 CM²

SLIGHTLY SHORTER BAKING TIME

with the same diameter. This means that going from a round tin to a square tin (keeping the amount of batter constant) creates a thinner layer of batter that requires a shorter baking time.

+ **Dark-coloured tins absorb and emit more heat than lighter ones** because of 'radiation heat transfer'. This means that a darker cake tin can result in a slightly shorter baking time and more caramelised edges, as well as slightly more doming. Usually, this isn't a huge effect, but it's still worth keeping in mind. (It's also the reason why I prefer lighter anodised aluminium cake tins for layer cakes and darker loaf tins for bakes such as banana bread.)

In the vast majority of cases, bake your cakes in metal tins, rather than ceramic or glass baking dishes. Metal tins are excellent conductors of heat and allow your cakes to bake evenly and quickly. All recipes in this chapter are perfected for such cake tins – if you, therefore, for whatever reason decide to use a cake tin made from a different material, you can expect the baking times to deviate from those listed in the recipes.

Of course, baking the cake perfectly doesn't help you all that much if said cake stubbornly sticks to the tin and refuses to budge. Even if a tin claims to be non-stick (and any number of other magical things that should apparently make your cake almost jump out of there), I recommend that you take steps to ensure you can get it out in one beautiful, intact piece. For round and square cake tins and loaf tins, grease generously with butter and line with baking paper. For bundt tins, grease generously with butter and dust with some almond flour (for cakes light in colour) or cocoa powder (for chocolate cakes). I've found that the extra fat in almond flour and cocoa powder makes them much more effective releasing agents than gluten-free flours.

IS IT BAKED?

The perfect wet:dry ingredients' ratio, the reverse creaming method, all the almond flour in the world, the correct cake tin and even the addition of chocolate chips won't help you if you over- or underbake your cake. (And when even chocolate doesn't help, you know it's a serious issue indeed.)

Many problems you might encounter in cake baking have to do with an incorrect oven temperature and baking time. An underbaked cake will sink in the middle, have a dense texture and is, in nine out of ten cases, suitable for only the bin. An overbaked cake, on the other hand, can be just as unpleasant – especially in gluten-free baking, where a dry crumb is a baker's worst nightmare. Luckily, a slightly dry, overbaked cake can be at least partly saved by soaking the sponge with a simple syrup of water and sugar, or some warm milk or brewed coffee, depending on the other flavours present.

However, it's better to avoid the issue of over- and underbaking altogether. And it's easily done. First, use an oven thermometer to ensure that your oven is properly calibrated, and recalibrate it if necessary (and if possible). (For more on ovens, see page 25.)

Second, take note of the oven temperature and baking time in the recipe. While you should always follow the oven temperature listed, there is some leeway when it comes to the time. To determine the optimal time, and therefore when the cake is perfectly baked, we return to our best friend that we've talked about before, and will talk about again (many times): the toothpick test (see page 30).

For best results, do a toothpick test 4–5 minutes before the baking time runs out. If the toothpick (or skewer) comes out with uncooked or half-baked batter attached, bake the cake for a few minutes more and check again. Repeat until you get a clean toothpick, dotted with just a few stray crumbs. You've got yourself a cake that's neither over- nor underbaked.

That the outer appearance of a cake can indicate whether or not the cake is baked is a common misconception. Depending on the amount of sugar and other ingredients present, as well as the shape (especially

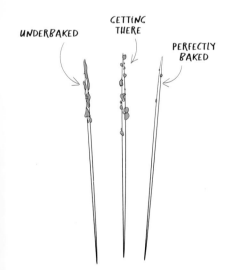

UNDERBAKED

GETTING THERE

PERFECTLY BAKED

the depth) of the cake tin and the consistency of the batter, some cakes start browning faster than others and/or take longer to bake. If your cake looks done, check it with a toothpick and, if it's not quite baked yet, cover it with a sheet of foil – shiny side up. This will reflect some of the radiative heat and slow down the rate of browning. And, you'll avoid an underbaked cake with a burnt top.

SCALING RECIPES UP AND DOWN

In general, you can scale all cake recipes up or down merely by multiplying all the ingredient quantities by a common number and keeping the ingredient ratios the same. (Any issues about scaling the raising agents differently from the rest of the ingredients become relevant only with industrial quantities.) The challenge, then, is to figure out what that number is, considering the size and shape of the suggested cake tin and the one you intend to use.

Now, I could bore you with a few pages of formulae derivations of how to get to that number… or I could just give you a simple, foolproof method to determine by how much to scale your recipe up or down, depending on the cake tin size and shape. Don't worry, I choose the second option. Take a look opposite for what you need.

All this simple mathematics relies on calculating the area of the different tins and assumes that the depth of the cake batter remains the same. If you're looking to change the thickness of the sponge for whatever reason, or you want to go from a simple cake tin to a loaf tin or (a much trickier case) to a bundt tin – I definitely don't recommend trying to calculate your way out of it! In such cases (or if you simply don't feel like working out equations to bake a cake), the easiest way to figure out how much batter to use is to fill the smaller of the tins with water and see how many tinfuls of water fit into the larger cake tin. It works every time.

When you do use the calculations opposite, however, it might happen that you end up with an inconvenient number by which to multiply your ingredients, giving you, for example, a result that means you have to use 4.37 eggs. If that's the case, recalculate the recipe to the nearest whole egg. (Now that's one sentence I've never thought I'd write!) When scaling recipes, eggs are always the deciding factor, as you really don't want to be measuring out ⅛ of one! In this example, you should recalculate all the ingredients in the recipe to use 4 eggs.

Changing the size of your cake will naturally influence the baking time, with larger cakes requiring a longer time. This is not such a simple calculation, and it's best to check the doneness of your cake with the toothpick test (see page 30).

Finally, whenever you've recalculated the recipe and successfully baked the (super-delicious and stunning) cake, make sure to note down the cake tin you used, the ingredient quantities you needed and the perfect baking time. The last thing you want to do is to have to recalculate the recipe every single time.

THE MATHEMATICS OF SCALING CAKES

Before you attempt to use the formulae, here's some nomenclature:
d = diameter, r = radius (half of the diameter), l = length, w = width,
q_A refers to a quantity to do with tin A, n = the number we're looking for.
In all of the below, we want to take a recipe written for tin A and adapt
it to tin B.

For scaling recipes between round cake tins: $n = (d_B/d_A)^2$
For scaling recipes between rectangular cake tins: $n = (l_B \times w_B)/(l_A \times w_A)$
For scaling recipes from a rectangular (A) and to a round (B) cake tin:
  $n = (3.14 \times r_B^2)/(l_A \times w_A)$
For scaling recipes from a round (A) and to a rectangular (B) cake tin:
  $n = (l_B \times w_B)/(3.14 \times r_A^2)$

Once you have the cake-tin measurements, just plug them in and let these formulae do the work for you. For example, if the recipe is for a 15cm round cake tin, but you want to use a 20cm round cake tin, the calculation is as follows:
$n = (20/15)^2 = 1.33^2 = 1.77$

Assuming the original recipe requires 1 egg, you'd now have to measure out an inconvenient 1.77 eggs. Therefore, you round up the value of n to 2. That is, to go from a 15cm to a 20cm round cake tin, you have to double all the ingredients in the recipe.

$$n = \left(d_B/d_A\right)^2$$

$$n = \left(l_B \times w_B\right)/\left(l_A \times w_A\right)$$

$$n = (3.14 \times r_B^2)/(l_A \times w_A)$$

$$n = \left(l_B \times w_B\right)/(3.14 \times r_A^2)$$

A NOTE ON BUTTERCREAMS + FROSTINGS

I believe in being generous with the buttercream, but also in keeping it not overly sweet. This has less to do with it being healthier (I mean, there's still more than enough sugar in there to tell you that this is 100% a treat) and far more to do with the fact that large amounts of sugar tend to overwhelm and mask any other flavours. That also guides my choice of the type of buttercream I use.

American buttercream is reserved for flavoured toppings, such as chocolate (see page 46) or strawberry (see page 108). That's because it usually relies on sugar to create the fluffy texture and prevent it from being too buttery in flavour (which can quickly turn greasy on the palate) – but ingredients such as chocolate and cocoa powder reduce the need for sugar.

Swiss meringue buttercream is my go-to topping where the main flavouring is less intense, such as vanilla (see page 50). I love this buttercream because it's nowhere near as sweet as American buttercream, while being fluffy and almost marshmallow-y in its texture, thanks to the meringue base. It also pipes and spreads beautifully.

That said, if you're a Swiss meringue buttercream novice, making it can be slightly daunting, what with making a cooked meringue first and then (rather counterintuitively) adding a whole lot of butter to it. The buttercream can also be slightly temperamental and it tends to go through a few different stages texture-wise, so it's easy to give up too quickly, simply because you didn't persist in mixing it for an extra minute or two. However, the good news is that Swiss meringue buttercream is pretty impossible to mess up if you've whisked up your meringue properly to stiff peaks – and as long as you don't panic. You can mostly solve any problems by adjusting the temperature of the buttercream, which should be somewhere around 22–24°C.

Here's a quick Swiss-meringue-buttercream troubleshooting list, including the three most common problems and how to solve them:

+ **It looks curdled.** Almost every single Swiss meringue buttercream will go through a stage where it looks curdled and split. Don't panic – just keep adding the butter according to the recipe and continue whisking the buttercream, it will eventually come together.

+ **It looks curdled after prolonged whisking.** If longer whisking didn't help, measure the temperature of the buttercream. If it's below 21°C, your butter was probably too cold to incorporate properly into the meringue. You need to heat it up. I do this by heating the outside of the bowl with a hairdryer while whisking. The buttercream should then lose the curdled look and turn smooth and fluffy. (Or, you can heat up a small portion of the buttercream in the microwave and add it back to the rest; or place the bowl over a pan of simmering water for a few seconds to gently heat the whole lot.)

+ **It looks like a soup.** Again, this is a problem of temperature. This time your buttercream is probably too warm – above 25°C – because you didn't cool the meringue down to room temperature before adding the butter. Consequently, the butter has melted within the meringue, resulting in a soupy, runny mixture that doesn't hold its shape. To solve this, just pop the buttercream, bowl and all, into the fridge for 15–30 minutes until it only just begins to firm up around the edges. Then whisk it up again until you get a smooth, fluffy result.

Double- or whipping-cream-based frostings are great when you want to incorporate dairy products like mascarpone or cream cheese, which don't hold their shape very well if whipped with sugar on their own. Instead of adding a small mountain of icing sugar in the hope that will stabilise the frosting, or making what is essentially a buttercream with only the teensiest amount of mascarpone or cream cheese added, I like to use a mixture with double cream (up to a 1:1 ratio of double cream to mascarpone or cream cheese). The results require less sugar and are great for spreading on to cakes or piping on to cupcakes, as the frostings hold their shape really well.

SUPER-MOIST CHOCOLATE CAKE

SERVES 12–14
PREP TIME 1 hour
BAKE TIME 35 mins

The initial version of this cake relied on only cocoa powder for the chocolate flavour. I won't lie – it felt like cheating. As delicious the cake was, cocoa powder simply doesn't match actual melted chocolate when it comes to giving a cake that deep, luxurious chocolate flavour. In this improved version, I've used both melted chocolate and cocoa powder in the sponges as well as in the frosting – and all of a sudden, it's become every chocoholic's dream come true. Because chocolate has a slightly acidic pH, it helps boost the activity of the raising agents and helps makes the sponges even more tender and delicate, without venturing into crumbly, falling-apart territory. Don't be scared by the addition of hot water to the cake batter – together with the milk, it helps keep the cake incredibly moist, so that it simply melts in your mouth.

CHOCOLATE SPONGES

280g dark chocolate (60–70% cocoa solids), chopped
200g unsalted butter
250g caster sugar
4 eggs, at room temperature
160g gluten-free flour blend
80g almond flour
40g Dutch processed cocoa powder
2 teaspoons baking powder
1 teaspoon bicarbonate of soda
½ teaspoon xanthan gum
½ teaspoon salt
120g hot water
120g whole milk, warmed

CHOCOLATE BUTTERCREAM

600g unsalted butter, softened
400g icing sugar
100g Dutch processed cocoa powder
½ teaspoon salt
300g dark chocolate (60–70% cocoa solids), melted and cooled

CHOCOLATE SPONGES

+ Adjust the oven shelf to the middle position, preheat the oven to 180°C and line two 20cm round cake tins with baking paper.
+ In a heatproof bowl above a pan of simmering water, melt the dark chocolate and butter together. Set aside to cool until warm, then add the sugar and eggs, and mix well.
+ Sift in the gluten-free flour blend, almond flour, cocoa powder, baking powder, bicarbonate of soda, xanthan gum and salt. Mix well until you get a smooth batter with no flour clumps.
+ Add the hot water and milk. Whisk well until combined.
+ Divide the batter equally between the prepared cake tins and bake for about 35–40 minutes or until an inserted toothpick comes out clean.
+ Allow to cool in the cake tins for 10 minutes, then turn out on to a wire rack to cool completely.

CHOCOLATE BUTTERCREAM

+ In a stand mixer with the paddle attachment or using a hand mixer fitted with the double beaters, beat the butter for 2–3 minutes. Sift in the icing sugar and beat for a further 5 minutes until pale and fluffy.
+ Sift in the cocoa powder and salt, and beat until evenly distributed.
+ Finally, add the melted and cooled chocolate and beat until you get a rich, smooth chocolate buttercream.

ASSEMBLING THE CAKE

+ If the sponges are domed, level them using a sharp, serrated knife.
+ Place the bottom sponge layer on a cake stand and spread a generous layer of buttercream on top, but leave enough buttercream for the outside of the cake. Sandwich using the other sponge layer, turned so that its bottom faces upwards. This will give the cake a nice, flat surface for decoration.
+ Use the remaining chocolate buttercream to cover the outside of the cake. Decorate the cake by creating swirls of buttercream with an offset spatula or the back of a spoon.

STORAGE

3–4 days in an airtight container or wrapped in cling film in a cool, dry place or the fridge. If you keep the cake in the fridge, leave it out at room temperature for about 15 minutes before serving.

PUT A TWIST ON IT

Milk chocolate cake – In the sponges, replace the dark chocolate with milk chocolate (for best results, melt the milk chocolate and butter separately), reduce the sugar to 200g, increase the gluten-free flour blend to 240g, and omit the cocoa powder. In the buttercream, replace the dark chocolate with milk chocolate and reduce the cocoa powder to 40g (but add it slowly until you get the desired flavour and colour).

White chocolate cake – In the sponges, replace the dark chocolate with white chocolate and reduce the butter to 160g (for best results, melt the white chocolate and butter separately), reduce the sugar to 150g, increase the gluten-free flour blend to 240g, increase the almond flour to 120g, and omit the cocoa powder. In the buttercream, replace the dark chocolate with white chocolate and omit the cocoa powder.

PHOTOGRAPHY OVERLEAF ⟶

SIMPLE VANILLA CAKE

SERVES 12–14
PREP TIME 1 hour 30 mins
BAKE TIME 20 mins

This cake is all about balance – and answering the question of 'How do you mix just the right ingredients in just the right proportions to get a sponge that is buttery, rich, flavourful, light and fluffy all at once?' There are two main responses here: using the reverse creaming method (see pages 38–9) to get a smooth, velvety crumb, and adding an extra egg white for every egg to give the cake a lift without compromising the flavour (as increasing the raising agents might do).

VANILLA SPONGES

300g gluten-free flour blend
100g almond flour ①
375g caster sugar
4½ teaspoons baking powder
¾ teaspoon xanthan gum
½ teaspoon salt
255g unsalted butter, softened
3 eggs, at room temperature
3 egg whites, at room temperature
180g whole milk, at room
    temperature
1½ teaspoons vanilla bean paste
4½ teaspoons lemon juice ②
fresh fruit, to decorate (optional)

VANILLA SWISS MERINGUE BUTTERCREAM

8 egg whites
500g caster sugar
½ teaspoon cream of tartar
500g unsalted butter, softened
1½ teaspoons vanilla bean paste
½ teaspoon salt

VANILLA SPONGES

+ Adjust the oven shelf to the middle position, preheat the oven to 180°C and line three 20cm round cake tins with baking paper.
+ In a large bowl, sift together the gluten-free flour blend, almond flour, sugar, baking powder, xanthan gum and salt.
+ Add the butter and, using a stand mixer with the paddle attachment or a hand mixer fitted with the double beaters, work the butter into the dry ingredients until the texture resembles coarse breadcrumbs. ③
+ In a separate bowl, mix together the eggs, egg whites, milk, vanilla paste and lemon juice. Add the wet ingredients to the flour mixture in two or three batches, mixing well after each addition, until you get a smooth batter with no flour clumps.
+ Divide the batter equally between the prepared cake tins, smooth out the tops, and bake for about 20–24 minutes or until risen, golden brown on top and an inserted toothpick comes out clean.
+ Allow the sponges to cool in the cake tins for about 10 minutes, then turn out on to a wire rack to cool completely.

VANILLA SWISS MERINGUE BUTTERCREAM

+ Mix the egg whites, sugar and cream of tartar in a heatproof bowl over a pan of simmering water. Stir continuously until the meringue mixture reaches 65°C and the sugar has dissolved.
+ Remove the mixture from the heat, transfer it to a stand mixer with the whisk attachment, or use a hand mixer fitted with the double beaters, and whisk on medium–high speed for 5–7 minutes until greatly increased in volume and at room temperature. It's very important that the meringue is at room temperature before you move on to the next step.
+ Add the butter, 1–2 tablespoons at a time, while mixing continuously on medium speed. If using a stand mixer, I recommend switching to the paddle attachment for this step. ④
+ Continue until you've used up all the butter and the buttercream looks smooth and fluffy. Add the vanilla paste and salt, mixing briefly to incorporate them.

ASSEMBLING THE CAKE

+ If the sponges are domed, level them out using a sharp, serrated knife.
+ Place the bottom sponge on a cake stand and spread a generous layer of buttercream on top. Sandwich with a second sponge, followed by another layer of buttercream and the final sponge on top, turned so that its bottom faces upwards. This will give the cake a nice, flat surface.
+ Use the remaining buttercream to cover the outside of the cake. Decorate the cake by creating swirls of buttercream with an offset spatula or the back of a spoon. Top with pieces of fresh fruit (such as raspberries), if you wish.

STORAGE

3–4 days in an airtight container or wrapped in cling film in a cool, dry place or the fridge. If you keep the cake in the fridge, leave it out at room temperature for about 15 minutes before serving.

NOTES

① *Almond flour keeps the cake moist for days (see page 35). You can substitute it with an equal weight of gluten-free flour blend, but the sponges will dry out faster.*

② *The lemon juice reacts with the raising agents to help make the sponges even lighter and fluffier.*

③ *The reverse creaming method (see pages 38–9) gives the sponges a smooth, velvety crumb and an even aeration, while reducing the likelihood of a domed top.*

④ *Add the butter slowly, so that the amount of fat doesn't overwhelm the meringue. You're creating an emulsion, which is all about balance and timing. If the buttercream goes through a 'soupy' stage, where it's very runny and looks split, continue adding butter and whisking/ beating the mixture. It will come together once you've added all the butter. For troubleshooting tips on Swiss meringue buttercream, see page 44.*

PHOTOGRAPHY OVERLEAF ⟶

HAZELNUT + MILK CHOCOLATE CAKE

SERVES 10–12
PREP TIME 1 hour 30 mins
COOK TIME 5 mins
BAKE TIME 30 mins
CHILL TIME 30 mins

If I could eat only one cake for the rest of my life, it would be this one. It tastes like Nutella (without actually using any Nutella) in cake form, which makes it about 1,000 times better. The sponges are buttery and delicate and have an intense hazelnut aroma thanks to roasting the hazelnuts before grinding them. Then, there's the milk chocolate buttercream, which is – in one word – dangerous (in the could-eat-it-by-the-spoonful way). Put the sponges and the buttercream together and you've got yourself fireworks.

HAZELNUT SPONGES

210g hazelnuts
210g gluten-free flour blend
300g caster sugar
3 teaspoons baking powder
¾ teaspoon xanthan gum
½ teaspoon salt
165g unsalted butter, softened
3 eggs, at room temperature
240g whole milk, at room
    temperature
1 teaspoon vanilla bean paste

MILK CHOCOLATE BUTTERCREAM

400g unsalted butter, softened
250g icing sugar
250g milk chocolate (about 40%
    cocoa solids), melted and cooled
20–30g Dutch processed cocoa
    powder
½ teaspoon salt

DARK CHOCOLATE DRIP

60g dark chocolate (60–70%
    cocoa solids), chopped
90g double cream

HAZELNUT SPONGES

+ Adjust the oven shelf to the middle position, preheat the oven to 180°C and line three 15cm round cake tins with baking paper.
+ First, toast the hazelnuts in a frying pan over a medium–high heat for 5–7 minutes, stirring frequently, until their skins start loosening and they get deep brown spots. ① Tip them into a tea towel, gather up the edges and use it to rub off their skins. Cool the nuts completely, discard the loosened skins, then set aside 30g for decoration later. Finely grind the remainder in a food processor or blender.
+ In a large bowl, sift together the gluten-free flour blend, sugar, baking powder, xanthan gum and salt. Add the ground hazelnuts and whisk to combine.
+ Add the butter and, using a stand mixer with the paddle attachment or a hand mixer fitted with the double beaters, work the butter into the dry ingredients until the texture resembles coarse breadcrumbs. ②
+ In a separate bowl, mix together the eggs, milk and vanilla paste. Add the wet ingredients to the flour mixture in two or three batches, and mix well until you get a smooth cake batter with no flour clumps.
+ Divide the batter equally between the prepared cake tins, smooth out the tops, and bake for about 30–32 minutes or until risen, golden brown on top and an inserted toothpick comes out clean. Don't open the oven during the first 20 minutes of baking, as it can cause the sponges to collapse. ③
+ Allow the sponges to cool in the cake tins for about 10 minutes, then turn out on to a wire rack to cool completely.

MILK CHOCOLATE BUTTERCREAM

+ In a stand mixer with the paddle attachment or using a hand mixer fitted with the double beaters, beat the butter for 2–3 minutes. Sift in the icing sugar and beat for a further 5 minutes until pale and fluffy.
+ Add the melted and cooled milk chocolate, and beat until incorporated.
+ Finally, sift in 20g of the cocoa powder and the salt, and beat until evenly distributed. You can add more cocoa powder (up to a total of about 30g) if you want a richer flavour and a slightly darker colour.

DARK CHOCOLATE DRIP

+ Place the chopped dark chocolate in a heatproof bowl.
+ In a saucepan, bring the double cream just to the boil, then pour it over the chocolate. Allow to stand for 4–5 minutes, then stir together until smooth and glossy.
+ Set aside to cool slightly.

ASSEMBLING THE CAKE

+ If the sponges are domed, level them out using a sharp, serrated knife.
+ Place the bottom sponge on a cake stand and spread a generous layer of buttercream on top. Sandwich with a second sponge, followed by another layer of buttercream and the final sponge on top, turned so that its bottom faces upwards. This will give the cake a nice, flat surface for decoration.
+ Use most of the remaining milk chocolate buttercream to cover the outside of the cake (leaving some for decorating), smoothing out the frosting with a bench scraper or offset spatula.
+ Refrigerate the cake for 30–45 minutes until the buttercream feels cool and firm to the touch.
+ Roughly chop the reserved toasted hazelnuts. Once the cake has chilled, use these to decorate the bottom third of the cake to create a 'collar' around the base.
+ Check the consistency of the chocolate drip. It should be runny enough to create a drip, but not so hot as to melt the frosting. If it's too thick, reheat it in the microwave in 5-second bursts until just right.
+ Create the chocolate drip. Either, use a squeezy bottle to make a rim of chocolate drip around the edge of the cake first, then fill in the top of the cake; or pour the chocolate drip on top of the cake and use a spoon/offset spatula to gently guide it to drip down the sides. Smooth out the top with an offset spatula and allow the drip to firm up slightly (this should only take a few minutes).
+ Transfer the remaining buttercream to a piping bag (use your choice of nozzle) and decorate the cake with piped swirls of buttercream to finish.

STORAGE

3–4 days in an airtight container or wrapped in cling film in a cool, dry place or the fridge. If you keep the cake in the fridge, leave it out at room temperature for about 15 minutes before serving.

NOTES

① *Toasting the hazelnuts gives them a much more intense, aromatic flavour and also helps loosen their skins. The latter is mostly an aesthetic consideration, as skinned hazelnuts result in beautiful, golden brown sponges.*

② *The reverse creaming method (see pages 38–9) gives the sponges a smooth, velvety crumb and an even aeration, while reducing the likelihood of a domed top.*

③ *While collapsing cakes usually aren't a problem with the reverse creaming method, the large amount of ground hazelnuts makes these sponges more delicate and extra fluffy directly out of the oven – and therefore also more likely to collapse if you open the oven too early.*

PHOTOGRAPHY OVERLEAF ⟶

CARROT CAKE

SERVES 10–12
PREP TIME 45 mins
BAKE TIME 1 hour 30 mins

Carrot cake is one of those rare treats where you're actually happy that it contains vegetables. Coarsely grated carrots give the cake a moist, delicate crumb (coarsely grating is essential – finely grated carrots release too much moisture during baking), while a mixture of warming spices contributes a wonderfully aromatic flavour. The layer of cream-cheese frosting is generously thick, without being too sweet or overpowering thanks to using double cream instead of butter.

CARROT CAKE

260g gluten-free flour blend
320g light brown soft sugar
1 teaspoon baking powder
1 teaspoon bicarbonate of soda
½ teaspoon xanthan gum
2 teaspoons ground cinnamon
½ teaspoon ground mixed spice
½ teaspoon ground ginger
½ teaspoon salt
200g vegetable, sunflower or
    other neutral-tasting oil
4 eggs, at room temperature
300g carrots, coarsely grated
140g walnuts or pecans, chopped

CREAM-CHEESE FROSTING

200g double cream, chilled ①
100g icing sugar, sifted
200g full-fat cream cheese, chilled
½ teaspoon vanilla bean paste
¼ teaspoon salt

NOTES

① *Double cream makes a more stable cream-cheese frosting than you could achieve using butter, even without large amounts of icing sugar (see page 45).*

② *Keeping the batter thick, and low in wet ingredients, ensures that the moisture released by the carrots doesn't result in a dense, wet and/or gummy cake.*

CARROT CAKE

+   Adjust the oven shelf to the middle position, preheat the oven to 180°C and line one 20cm round cake or springform tin (at least 7cm deep) with baking paper.
+   In a large bowl, sift together the gluten-free flour blend, sugar, baking powder, bicarbonate of soda, xanthan gum, all the spices and the salt.
+   Add the oil and eggs, and mix to a smooth, thick cake batter. ②
+   Add the grated carrots. Set aside 20g of the chopped nuts for decoration, then add the remainder to the batter and mix until evenly distributed.
+   Transfer the batter into the prepared tin, smooth out the top, and bake for about 1 hour 30 minutes or until risen, deep golden brown on top and an inserted toothpick comes out clean. If the top of the cake starts browning too quickly, cover with foil (shiny side up) and bake until done.
+   Allow the sponge to cool in the cake tin for about 15 minutes, then remove from the tin and transfer it to a wire rack to cool completely.

CREAM-CHEESE FROSTING

+   Using a stand mixer with the whisk attachment or a hand mixer fitted with the double beaters, whip the double cream and icing sugar together until stiff peaks form.
+   In a separate bowl, whip the cream cheese until smooth. Add it to the whipped cream and whip for about 1 minute to stiff peaks.
+   Add the vanilla paste and salt, and mix until incorporated.
+   Spread the frosting evenly on top of the cooled cake, creating swirls with an offset spatula or the back of a spoon. Sprinkle with the reserved 20g of chopped nuts to finish.

STORAGE

3–4 days in a closed container in a cool, dry place.

LEMON DRIZZLE CAKE

SERVES 10
PREP TIME 30 mins
BAKE TIME 1 hour

Absolutely loaded with lemon zest, this cake couldn't be more lemony if it tried. It also boasts the most perfect, moist, delicate crumb that pairs wonderfully with the lemon icing – especially when the icing dries into a crackly, refreshing, sugary goodness. The cake isn't overwhelmingly sweet. Instead, it is fragrant, aromatic and impossible to resist.

LOAF CAKE
200g gluten-free flour blend
65g almond flour ①
200g caster sugar
3 teaspoons baking powder
½ teaspoon xanthan gum
¼ teaspoon salt
225g unsalted butter, softened
zest of 3 unwaxed lemons
4 eggs, at room temperature
120g whole milk, at room
    temperature
6 tablespoons lemon juice
1 teaspoon vanilla bean paste

LEMON ICING
120g icing sugar, sifted
7–8 teaspoons lemon juice

NOTES
① Almond flour keeps the cake moist for days (see page 35). You can substitute it with an equal weight of gluten-free flour blend if you like, but the cake will dry out faster.

② The reverse creaming method (see pages 38–9) gives the cake a smooth, velvety crumb.

LOAF CAKE
+ Adjust the oven shelf to the middle position, preheat the oven to 180°C and line a 900g loaf tin (23 x 13 x 7.5cm) with baking paper.
+ In a large bowl, sift together the gluten-free flour blend, almond flour, sugar, baking powder, xanthan gum and salt.
+ Add the butter and, using a stand mixer with the paddle attachment or a hand mixer fitted with the double beaters, work the butter into the dry ingredients until the texture resembles coarse breadcrumbs. ② Add the lemon zest and mix well.
+ In a separate bowl, mix together the eggs, milk, lemon juice and vanilla paste. Add the wet ingredients to the flour mixture in two or three batches, mixing well after each addition, until you get a smooth cake batter with no flour clumps.
+ Transfer the batter into the prepared loaf tin, smooth out the top, and bake for about 1 hour or until risen, golden brown on top and an inserted toothpick comes out clean. If it starts browning too quickly, cover with foil (shiny side up) and bake until done.
+ Allow the cake to cool in the loaf tin for about 10 minutes, then remove from the tin and transfer to a wire rack to cool completely.

LEMON ICING
+ Whisk the icing sugar and 6 teaspoons of the lemon juice together until you get a thick mixture. Add a further 1–2 teaspoons of lemon juice, mixing well, until you get a runny icing consistency that coats the back of a spoon.
+ Drizzle the lemon icing on top of the cooled cake, spreading it evenly with the back of a spoon and letting it drip down the sides.

STORAGE
3–4 days in a closed container in a cool, dry place.

PUT A TWIST ON IT
Orange drizzle cake – Simply swap the lemons for oranges.
Pink drizzle – For this cheerful and vibrant drizzle alternative, substitute part of the lemon juice with a raspberry reduction: cook 100g raspberries until softened, pass them through a sieve to remove the seeds and reduce the juices over a gentle heat until thickened, then mix with the icing sugar.

WHOLE ORANGE + CHOCOLATE MARBLED BUNDT CAKE

SERVES 14–16
PREP TIME 1 hour
COOK TIME 2 hours
BAKE TIME 1 hour 10 mins

BUNDT CAKE

2–3 unwaxed oranges (preferably organic)
240g unsalted butter, softened, plus extra for greasing
120g almond flour, plus extra for dusting
400g caster sugar
4 eggs, at room temperature
350g gluten-free flour blend
2 teaspoons baking powder
1 teaspoon bicarbonate of soda
1 teaspoon xanthan gum
½ teaspoon salt
60g full-fat plain yoghurt or soured cream, at room temperature ①
100g dark chocolate (60–70% cocoa solids), melted and cooled
20g Dutch processed cocoa powder

CHOCOLATE GLAZE

100g dark chocolate (60–70% cocoa solids), chopped
135g double cream

I'm sure you will agree with me when I say: there are few flavour combinations as spectacular as chocolate and orange. To take this cake beyond just using the orange zest and juice, I have slowly simmered whole oranges for 2 hours before puréeing them and incorporating them into the cake. This gives the cake a deep, elegant flavour and an incredibly moist, rich crumb. It isn't light and fluffy – think dense and buttery, without being unpleasantly heavy. The swirl of chocolate batter and the chocolate ganache glaze complete a bundt cake intended truly to take your breath away.

BUNDT CAKE

+ Thoroughly wash the oranges, place them in a large saucepan filled with water and simmer for about 2 hours until softened. Take them out of the water and allow to cool. Cut away the tops and bottoms, cut into quarters and remove any seeds. Purée the orange quarters in a food processor or blender until smooth. Set aside until needed.

+ Adjust the oven shelf to the lower-middle position and preheat the oven to 180°C. Butter a large 12-cup bundt tin (about 26cm wide and 11cm deep) and dust it with almond flour.

+ Using a stand mixer with the paddle attachment or a hand mixer fitted with the double beaters, cream the butter and sugar together for about 5 minutes until pale and fluffy. Add the eggs, one at a time, mixing well after each addition. Scrape down the sides and bottom of the bowl from time to time to prevent any patches of unmixed butter.

+ Sift together the gluten-free flour blend, almond flour, baking powder, bicarbonate of soda, xanthan gum and salt. Add ⅓ of this flour mixture to the butter–egg mixture, mixing well until fully incorporated.

+ Add about 350g of the orange purée and mix well. Don't worry if the mixture separates or appears curdled (unlikely, but it can because of the large amount of moisture in the purée) – the next addition of the dry ingredients will make the batter perfectly smooth and creamy.

+ Add the second ⅓ of the flour mixture, the yoghurt or soured cream and then the remaining flour mixture, mixing well after each addition, to give a smooth, fluffy and fairly thick batter with no flour clumps. Again, scrape down the sides of the bowl from time to time.

+ Transfer about ⅓ of the batter (about 575g) into a separate bowl and add the melted chocolate and cocoa powder. Mix well until the ingredients are fully combined.

+ Spoon ⅓ of the plain orange batter into the bundt tin. Spoon ½ of the chocolate batter on top, then cover with ⅓ of the plain batter, followed by the rest of the chocolate batter and finish with a final layer of the remaining plain batter. Tap the bundt tin gently on the work surface to help the batter settle into the tin grooves. Use a knife or skewer to swirl the batter around all the way to the bottom. Smooth out the top.
+ Bake for about 1 hour 10 minutes or until an inserted toothpick comes out clean or with a few moist crumbs attached. If the top of the cake starts browning too quickly, cover with foil (shiny side up) and bake until done.
+ Remove from the oven and allow to cool in the tin for 20–30 minutes, then carefully turn out on to a wire rack to cool completely.

CHOCOLATE GLAZE
+ Place the chopped dark chocolate into a heatproof bowl.
+ In a saucepan, bring the double cream just to the boil, then pour it over the chocolate. Allow to stand for 2–3 minutes, then stir together until smooth and glossy.
+ Pour the glaze over the cooled bundt cake, allowing it to run down the sides.
+ Allow to set at room temperature for at least 30 minutes before serving.

STORAGE
3–4 days in a closed container or covered in cling film in a cool, dry place.

NOTE
① *The slight acidity of the yoghurt or soured cream gives the cake an extra freshness (as well as giving a boost to the raising agents), making the cake lighter both in flavour and in texture.*

PHOTOGRAPHY OVERLEAF ⟶

RASPBERRY TRAYBAKE

SERVES 12–16
PREP TIME 45 mins
BAKE TIME 50 mins
COOK TIME 5 mins

There's a cosy charm to a simple traybake. It's easy to prepare and doesn't require any fancy decorating work and yet it can be absolutely bursting with flavour. This one has a rich, buttery, lemon-zest-scented sponge dotted with raspberries that add colour and tangy freshness. The mascarpone frosting is smooth and velvety, swirled through with a raspberry reduction that ties this cake together.

TRAYBAKE

300g gluten-free flour blend
100g almond flour ①
300g caster sugar
4½ teaspoons baking powder
¾ teaspoon xanthan gum
½ teaspoon salt
340g unsalted butter, softened
zest of 2 unwaxed lemons
6 eggs, at room temperature
190g whole milk, at room
   temperature
1 tablespoon lemon juice
1 teaspoon vanilla bean paste
350g frozen raspberries ②

MASCARPONE RASPBERRY FROSTING

250g fresh or frozen raspberries
300g double cream, chilled
100–150g icing sugar, sifted
225g mascarpone cheese, chilled
1 teaspoon vanilla bean paste

TRAYBAKE

+ Adjust the oven shelf to the middle position, preheat the oven to 180°C and line a 23 x 33cm baking tray with baking paper.
+ In a large bowl, sift together the gluten-free flour blend, almond flour, sugar, baking powder, xanthan gum and salt.
+ Add the butter and, using a stand mixer with the paddle attachment or a hand mixer fitted with the double beaters, work the butter into the dry ingredients until the texture resembles coarse breadcrumbs. Add the lemon zest and mix until evenly distributed.
+ In a separate bowl, mix together the eggs, milk, lemon juice and vanilla paste. Add the wet ingredients to the flour mixture in two or three batches, mixing well after each addition, until you get a smooth cake batter with no flour clumps.
+ Transfer half the batter to the prepared baking tray, smooth it out into an even layer and scatter half the raspberries on top. Cover with the remaining batter, smooth out the top and sprinkle with the remaining raspberries. ③
+ Bake for about 50–55 minutes or until risen, golden brown on top and an inserted toothpick comes out clean. If the cake starts browning too quickly, cover with foil (shiny side up) and bake until done.
+ Depending on how you want to serve the cake, you can cool it completely in the baking tray, or leave it in the baking tray for 10 minutes, then remove it and transfer it to a wire rack to cool completely.

MASCARPONE RASPBERRY FROSTING

+ First, prepare the raspberry reduction by cooking the raspberries in a saucepan over a medium–high heat, stirring frequently, until the berries have softened and they've released their juices.
+ Pass the contents of the pan through a sieve into a jug or bowl to remove the seeds and skins.
+ Return the juices to the pan and cook over a medium–high heat, stirring frequently, for a further 3–5 minutes until thickened and viscous, but not quite jam-like. Set aside to cool until needed.
+ Using a stand mixer with the whisk attachment or a hand mixer fitted with the double beaters, whip together the double cream and icing sugar until soft peaks form.
+ In a separate bowl, whip the mascarpone until loose and smooth, then fold it into the sweet whipped cream along with the vanilla paste. The frosting at this point should be easily spreadable and form soft peaks. If it's too loose, whip it briefly to thicken slightly.
+ Spread the frosting on top of the cooled traybake and swirl through the cooled raspberry reduction. Cut into portions and serve.

STORAGE

3–4 days in a closed container or covered in cling film in a cool, dry place.

PUT A TWIST ON IT

You can use pretty much any other berries to make this traybake, from blueberries and blackberries to strawberries.

NOTES

① *Almond flour keeps the cake moist for days (see page 35). You can substitute it with an equal weight of gluten-free flour blend, but the cake will dry out faster.*

② *Frozen raspberries firm up the surrounding cake batter (by cooling down and setting the butter in the batter), which makes them less likely than fresh raspberries to sink to the bottom of the cake.*

③ *Scattering the raspberries in two batches ensures an even distribution of berries throughout the cake.*

PHOTOGRAPHY OVERLEAF ⟶

VANILLA + RASPBERRY SWISS ROLL

SERVES 10–12
PREP TIME 45 mins
BAKE TIME 10 mins
COOK TIME 5 mins

VANILLA SPONGE

sunflower, vegetable or other
   neutral-tasting oil, for greasing
4 eggs, at room temperature
150g caster sugar
½ teaspoon vanilla bean paste
60g unsalted butter, melted
110g gluten-free flour blend
¾ teaspoon xanthan gum ①
½ teaspoon baking powder
½ teaspoon bicarbonate of soda
¼ teaspoon salt

RASPBERRY WHIPPED CREAM

250g fresh or frozen raspberries
250g double cream, chilled
75–100g icing sugar, plus extra
   for dusting

Making a good Swiss roll comes down to just a few key points: knowing what consistency you're looking for when whisking the eggs with sugar, being gentle when folding in other ingredients to preserve the precious air bubbles in the batter, not overbaking and therefore drying out the sponge, and avoiding that dreadful (and traditional) roll-cool-unroll-fill-roll method (see Note #3), which is frequently the reason why Swiss rolls crack.

VANILLA SPONGE

+ Adjust the oven shelf to the middle position, preheat the oven to 180°C and line a 25 x 35cm shallow baking tray with baking paper. Lightly grease the baking paper and sides of the tin with oil (or spray it with a non-stick baking spray). ②

+ Using a stand mixer with the whisk attachment or a hand mixer fitted with the double beaters, whisk together the eggs and sugar for about 5–7 minutes on a medium–high speed until pale, thick, fluffy and about tripled in volume. The mixture should briefly pile on top of itself as it falls off the whisk.

+ Add the vanilla and melted butter and whisk for 10–20 seconds until fully incorporated.

+ Sift the gluten-free flour blend, xanthan gum, baking powder, bicarbonate of soda and salt into the wet ingredients, and whisk for 10–20 seconds until no flour clumps remain. Scrape down the sides and bottom of the bowl to prevent any unmixed patches.

+ Transfer the batter to the prepared baking tray and smooth it out into an even layer. Bake for about 10–12 minutes or until an inserted toothpick comes out clean and the sponge feels springy to the touch.

+ Once baked, immediately cover the baking tray tightly with foil. Allow to cool until room temperature or about 22°C (if you cool it too much the sponge can crack on rolling, while not cooling it enough can melt the filling). ③ While the sponge is cooling, prepare the filling.

RASPBERRY WHIPPED CREAM

+ First, prepare the raspberry reduction by cooking the raspberries in a saucepan over a medium–high heat until the berries have softened and they've released their juices.
+ Pass the contents of the pan through a sieve into a jug or bowl to remove the seeds and skins.
+ Return the juices to the pan and cook over a medium–high heat, stirring frequently, for a further 3–5 minutes until thickened and viscous, but not quite jam-like. Set aside to cool completely.
+ Whip the double cream and icing sugar to stiff peaks, then fold in the raspberry reduction – either until fully combined into a pale pink cream, or partially, so that you get streaks of white, pink and red.

ASSEMBLING THE SWISS ROLL

+ Remove the foil from the cooled sponge, loosen the sponge from the edges of the tray with an offset spatula and spread the filling on top in a smooth, even layer, leaving a 1–2cm edge free on one of the short edges (this is where you will start rolling up the sponge).
+ Turn the sponge so that the filling-free short edge is closest to you. Using the baking paper underneath to help you, lift the edge of the sponge and gently fold it over itself to start the roll.
+ Gently lift the baking paper to continue the roll all the way to the end – the baking paper should easily peel away from the sponge.
+ Trim the ends of the rolled cake and dust with icing sugar to serve.

STORAGE

3–4 days in an airtight container or wrapped in cling film in a cool, dry place or the fridge. If you keep the cake in the fridge, leave it out at room temperature for about 15 minutes before serving. (Note that because the cream gets its colour from a natural raspberry reduction, it can oxidise slightly on contact with air, getting a slightly violet tinge. This doesn't affect the flavour, only the appearance.)

PUT A TWIST ON IT

You can use the sponge recipe as the basis for numerous other Swiss rolls by adding flavourings and spices, or even food colourings to create patterned roulades. Or, you could fill the plain vanilla sponge with a jam of choice to make a jam roll.

NOTES

① *The slightly larger amount of xanthan gum (considering the amount of flour) ensures that the sponge stays flexible enough to roll without cracking but still without being rubbery.*

② *As the baking paper needs to release smoothly from the sponge during rolling, it's very important to grease it lightly.*

③ *Stella Parks from Serious Eats pioneered this amazing, foolproof method of keeping the sponge flexible enough to roll without cracking. The traditional method is to roll the hot sponge, let it cool, then unroll, fill and re-roll, but that puts too much strain on this delicate sponge. Stella's method uses the steam trapped by the foil to keep the sponge pliable.*

PHOTOGRAPHY OVERLEAF ⟶

CHOCOLATE SWISS ROLL

SERVES 10–12
PREP TIME 1 hour
BAKE TIME 10 mins

The perfect swirl of moist chocolate sponge and mascarpone whipped-cream frosting, drizzled with a luscious chocolate ganache glaze. In the world of cakes, the humble Swiss roll is often overlooked for being old fashioned and not worth the effort. But, when you take a bite and the chocolate sponge simply melts in your mouth, and the combination of creamy mascarpone filling and chocolate glaze sets off fireworks on your taste buds – well, you realise that's not the case at all.

CHOCOLATE SPONGE

sunflower, vegetable or other
  neutral-tasting oil, for greasing
60g dark chocolate (60–70% cocoa
  solids), chopped
60g unsalted butter
4 eggs, at room temperature
150g caster sugar
90g gluten-free flour blend
20g Dutch processed cocoa
  powder
¾ teaspoon xanthan gum ①
½ teaspoon baking powder ②
¼ teaspoon salt

MASCARPONE WHIPPED CREAM

240g double cream, chilled
50g icing sugar, sifted
60g mascarpone cheese, chilled

CHOCOLATE GLAZE

120g dark chocolate (about 60%
  cocoa solids), chopped
180g double cream

CHOCOLATE SPONGE

+ Adjust the oven shelf to the middle position, preheat the oven to 180°C and line a 25 x 35cm shallow baking tray with baking paper. Lightly grease the baking paper and sides of the tin with oil (or spray it with a non-stick baking spray). ③
+ In a heatproof bowl above a pan of simmering water, melt the chocolate and butter together until smooth and glossy. Set aside to cool slightly.
+ Using a stand mixer with the whisk attachment or a hand mixer fitted with the double beaters, whisk together the eggs and sugar for about 5–7 minutes on medium–high speed until pale, thick, fluffy and about tripled in volume. The mixture should briefly pile on top of itself as it falls off the whisk.
+ Add the melted chocolate and whisk for 10–20 seconds until fully incorporated.
+ Sift the gluten-free flour blend, cocoa powder, xanthan gum, baking powder and salt into the chocolate mixture, and whisk for 10–20 seconds until no flour clumps remain. Scrape down the sides and bottom of the bowl to prevent any unmixed patches.
+ Transfer the batter to the prepared baking tray and smooth out into an even layer. Bake for about 10–12 minutes or until an inserted toothpick comes out clean and the sponge feels springy to the touch.
+ Once baked, immediately cover the baking tray tightly with foil. Allow the sponge to cool until room temperature or about 22°C (if you cool it too much, it can crack on rolling, while not cooling it enough can melt the filling). ④ While the sponge is cooling, prepare the filling and glaze.

MASCARPONE WHIPPED CREAM

+ Using a stand mixer with the whisk attachment or a hand mixer fitted with the double beaters, whip together the double cream and icing sugar until stiff peaks only just begin to form. ⑤
+ In a separate bowl, lightly whip the mascarpone until smooth, then fold it into the whipped cream until you get a smooth filling that spreads easily but will also hold its shape.

ASSEMBLING THE SWISS ROLL

+ Remove the foil from the cooled sponge, loosen the sponge from the edges of the tray with an offset spatula and spread the filling on top in a smooth, even layer, leaving a 1–2cm edge free on one of the short edges (this is where you will start rolling up the sponge).
+ Turn the sponge so that the filling-free short edge is closest to you. Using the baking paper underneath to help you, lift the edge of the sponge and gently fold it over itself to start the roll.
+ Gently lift the baking paper to continue the roll all the way to the end – the baking paper should easily peel away from the sponge. Set the rolled cake aside while you prepare the chocolate glaze.

CHOCOLATE GLAZE

+ Place the chopped dark chocolate in a heatproof bowl.
+ In a saucepan, bring the double cream just to the boil, then pour it over the chocolate. Allow to stand for 2–3 minutes, then stir together until smooth and glossy.
+ Pour the glaze over the Swiss roll, allowing it to run down the sides.
+ Allow the glaze to set at room temperature for at least 30 minutes before serving.

STORAGE

3–4 days in an airtight container or wrapped in cling film in a cool, dry place or the fridge. If you keep the Swiss roll in the fridge, leave it out at room temperature for about 15 minutes before serving.

PHOTOGRAPHY OVERLEAF ⟶

NOTES

① *The slightly larger amount of xanthan gum (considering the amount of flour) ensures that the sponge stays flexible enough to roll without cracking but still without being rubbery.*

② *The melted chocolate in the batter is slightly acidic, giving the raising agents a boost. Consequently, you need a smaller quantity of raising agent than you do for the vanilla version of the Swiss roll (see page 70).*

③ *As the baking paper needs to release smoothly from the sponge during rolling, it's very important to grease it lightly.*

④ *Stella Parks from Serious Eats pioneered this amazing, foolproof method of keeping the sponge flexible enough to roll without cracking. The traditional method is to roll the hot sponge, let it cool, then unroll, fill and re-roll, but it puts too much strain on this delicate sponge. Stella's method uses the steam trapped by the foil to keep the sponge pliable.*

⑤ *Use the low–medium setting on your mixer to avoid very large air bubbles and overwhipping. If you overshoot and your whipped cream appears stiff and almost curdled, gently stir in more cream (without whipping) to fix.*

CHOCOLATE LAVA CAKES

MAKES 4
PREP TIME 45 mins
CHILL TIME 30 mins
BAKE TIME 12 mins

CHOCOLATE GANACHE
60g dark chocolate (60–70% cocoa
solids), chopped
90g double cream

LAVA CAKES
60g unsalted butter, plus extra for
greasing
20g Dutch processed cocoa
powder, plus extra for dusting
100g dark chocolate (60–70%
cocoa solids), chopped
100g light brown soft sugar ①
2 eggs, at room temperature
20g gluten-free flour blend
¼ teaspoon xanthan gum
¼ teaspoon salt
icing sugar, for dusting
double cream or vanilla ice cream
and strawberries, to serve

NOTES
① *Using light brown soft sugar,
keeping the amount of flour low,
and using a relatively large
amount of chocolate and butter,
keeps the sponge moist and
prevents it from drying out.*

② *To prepare the lava cakes in
advance, follow the recipe to
this point, then chill until needed
(but ideally bake them the same
day). Bake them directly out
of the fridge, but increase the
baking time by about 2 minutes
(to a total of 14 minutes).*

Who doesn't love a chocolate lava cake, with its seductively molten centre, especially when served with a generous scoop of ice cream? Instead of relying on underbaked cake batter for the 'lava' part, this recipe uses chilled chocolate ganache. Not only does this eliminate any fear of uncooked eggs in the batter, it's much richer in flavour and more luxurious in texture. Using a ganache centre also relieves some of the need to time the baking down to the second – the ganache will still be runny even if you bake it for an extra minute or so.

CHOCOLATE GANACHE
+ Place the chopped dark chocolate into a heatproof bowl.
+ In a saucepan, heat the cream until it just comes to the boil, then pour it over the chocolate. Allow to stand for 2–3 minutes, then stir well to a smooth, glossy ganache.
+ Refrigerate for about 30 minutes until firmed up but still scoopable.

LAVA CAKES
+ Adjust the oven shelf to the middle position, preheat the oven to 220°C and generously butter 4 ramekins and dust them with cocoa powder. Tap out any excess cocoa powder and set aside.
+ In a heatproof bowl over a pan of simmering water, melt the dark chocolate and butter together. Set aside to cool until warm.
+ Using a stand mixer with the whisk attachment or a hand mixer fitted with the double beaters, whisk the sugar and eggs on medium–high speed for 5–7 minutes until pale, thick, fluffy and tripled in volume. The mixture should briefly pile on top of itself as it falls off the whisk.
+ Fold in the melted chocolate until only just incorporated. Sift the gluten-free flour blend, cocoa powder, xanthan gum and salt into the wet ingredients, and gently fold in the dry using a spatula until no flour clumps remain. Scrape down the sides and bottom of the bowl from time to time to prevent any unmixed patches.
+ Divide half the batter between the 4 prepared ramekins. Place about 1½ tablespoons of the cooled ganache in the centre of each partially filled ramekin. Divide the remaining batter between the ramekins, completely covering the ganache centre. ②
+ Place the ramekins on a baking sheet or tray, and bake for about 12 minutes or until the cake tops have risen and feel firm to the touch, but the cakes wobble slightly in the centre when shaken. Allow to cool for 20–30 seconds, then invert on to serving plates. Dust with icing sugar and serve with cream or ice cream, and a few strawberries.

PUT A TWIST ON IT
Instead of chocolate ganache, fill the lava cakes with salted caramel or chocolate caramel ganache (just pour hot caramel over chopped chocolate).

CUPCAKES

+

MUFFINS

For the longest time, I had a love–hate relationship with cupcakes and muffins. Time after time, my cupcakes came out of the oven comically misshapen (at times so misshapen, in fact, that they passed from muffin straight into Yorkshire pudding territory), and my muffins all flat and sad-looking. I called it 'The Big Muffin Conspiracy'. It took me weeks of experimenting until I finally arrived at a basic vanilla cupcake recipe I was actually happy with. It was flat, with only a gentle roundness to its top, perfectly buttery, and had the most amazing, delicate, melt-in-the-mouth crumb. Crucially, it stayed that way for days afterwards – no dry, dense next-day cupcakes in sight.

The muffin counterpart, a reliable vanilla muffin recipe that produced a gorgeous bakery-style domed top and a fluffy, moist crumb, followed soon afterwards – because as I was learning about all the factors that make cupcakes flat, so I was discovering the things that make muffins rise to glorious heights.

Nowadays, The Big Muffin Conspiracy is a thing of the past – thankfully. Instead, there are muffins that taste like apple pie and cupcakes stuffed with chocolate fudge sauce – and peals of disbelieving laughter at that strange period when such delicious wonders seemed impossible.

SUPPOSED TO BE A CUPCAKE

ALSO A... CUPCAKE?
(DISGUISED AS A YORKSHIRE PUDDING)

MEANT TO BE A MUFFIN... SIGH

A *VERY* CONFUSED MUFFIN

THE TYPES OF CUPCAKE + MUFFIN IN THIS CHAPTER

I've chosen all the recipes in this chapter not only for their deliciousness, but also because they're one of the following:

+ basic, fundamental recipes from which many others follow,
+ recipes that introduce a particularly interesting ingredient or method that affects how the cupcakes and muffins behave both in and out of the oven.

Okay, I admit that stuffing the chocolate cupcakes with chocolate fudge filling (see page 88) isn't strictly necessary for your insight into the world of gluten-free cupcakes, but it's my life's mission to stuff as many cupcakes with as many delicious things as possible. Which is why four out of the seven cupcakes in this book are stuffed – with everything from coffee caramel sauce to lemon curd. You're very welcome.

The Big Muffin Conspiracy. Once upon a time, my cupcakes looked like Yorkshire puddings and my muffins were flat and rather sad. Thankfully, the conspiracy is now a thing of the past and my cupcakes and muffins are instead like those illustrated opposite, every time.

THE ROLE OF GLUTEN

Much as with cakes, gluten isn't a major player in the world of cupcakes and muffins. With wheat-based baking, gluten can, in fact, be a hindrance to getting a perfectly delicate, melt-in-the-mouth cupcake crumb if you overmix the batter.

Overmixing isn't much of an issue with gluten-free cupcakes and muffins, and it's easy to replicate any of the positive effects of gluten (such as retaining moisture and giving structure) by tweaking the wet:dry ingredients' ratio and adding a pinch of xanthan gum.

MUFFINS AREN'T JUST CUPCAKES WITHOUT BUTTERCREAM – AND OTHER DIFFERENCES

In baking, just like in science, definitions are important. Before we start looking at how you make the perfect gluten-free cupcakes and muffins, let's have a chat about what makes a cupcake a cupcake and a muffin a muffin. Let's set the stage, so to speak, for the deliciousness to come.

Cupcakes. Their defining feature is without doubt an almost flat, only gently rounded top, which creates the perfect surface for a small mountain of beautifully piped buttercream. They should be buttery, with a tender, melt-in-the-mouth crumb. They should definitely be on the sweeter side, but should take into account the buttercream finish – the last thing anyone wants is a sickly sweet cupcake where the only thing you taste is sugar. Still, cupcakes are firmly on the dessert side of things.

Muffins. On the other hand, muffins are only hinting at being a dessert – and thank goodness: who doesn't love a muffin for breakfast or as a snack? Muffins aren't as sweet as cupcakes – they're more about flavours and textures. They're almost bready in terms of their crumb, which is much more robust and hearty, and moist enough to still be pleasant to eat the following day (or even up to four days later if briefly popped into the microwave to reheat). Then, of course, there's the muffin top. Proudly tall, gloriously domed, with a gorgeous golden-brown tan, and preferably extending beyond the width of the muffin case or the muffin-tin hole, into what many call 'bakery-style muffins'.

To make a cupcake a cupcake and a muffin a muffin, then, the two have significantly different ingredient ratios, batter consistencies and baking temperatures – all reflected in the recipes in this chapter.

THE IDEAL
(VANILLA) CUPCAKE

LOADED WITH
BUTTERCREAM

MINIMAL
CARAMELISATION

FLAT, ONLY GENTLY
ROUNDED TOP

BUTTERY, FLUFFY,
TENDER CRUMB

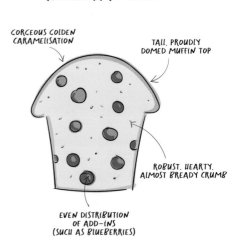

THE IDEAL
(BLUEBERRY) MUFFIN

GORGEOUS GOLDEN
CARAMELISATION

TALL, PROUDLY
DOMED MUFFIN TOP

ROBUST, HEARTY,
ALMOST BREADY CRUMB

EVEN DISTRIBUTION
OF ADD-INS
(SUCH AS BLUEBERRIES)

5 REASONS CUPCAKES ARE FLAT + MUFFINS ARE DOMED

Originally, I planned to write two sections here – one about how you get flat cupcakes and one about how to get tall, domed muffins. It quickly became clear that I was pretty much writing the same thing twice: factors that make cupcakes flat in opposite make muffins domed, unsurprisingly enough. Instead, here's the unified overview of the five most important things to consider when baking cupcakes and muffins, also summarised in Table 4 (see page 86).

1. Batter consistency. As a general rule, muffin batter should be significantly thicker than cupcake batter. While cupcake batter spreads evenly once scooped into the cupcake liners, muffin batter retains the shape of the scoop, seemingly having a predisposition towards forming a domed top. The science behind this predisposition revolves around the density and viscosity of the batter, and their impact on the rate of heat transfer – and, truth be told, we don't (yet) fully understand it. The crucial takeaway is that, for best results, you should keep your muffin batter thick and your cupcake batter loose.

2. Filling the liners. When it comes to filling the paper liners (muffin cases, cupcake cases and so on), I don't subscribe to the 'fill-them-halfway' idea. With cupcakes, I like to fill the liners about ¾ full, which ensures that the cupcake sides only just peak over the rim once baked – I'm not a fan of cupcakes that disappear below the rim and hide away, so that the only thing you can see is the icing.

With muffins, you need to fill the paper liners to the brim, possibly with a small mound a few millimetres above the rim – especially if you're aiming for bakery-style, jumbo muffins, where the top extends beyond the diameter of the liner into that iconic mushroom-like shape.

In all the recipes, the quantities and baking times are for 12-hole muffin or cupcake tins with 120ml holes, and paper liners that are 7cm diameter at the top, 5cm diameter at the bottom and 4cm high.

3. Oven temperature. Bake muffins at a relatively high temperature of 190°C, which makes the edges of the batter fully bake through and set before the centres are done. Consequently, the continued evolution of carbon dioxide thanks to the raising agents (and the water vapour from water evaporation) means that the centres keep on rising while the edges stay put. Voilà! We've got a muffin top. While many wheat-based recipes start muffins off at temperatures above 200°C, I don't recommend this for gluten-free muffins because gluten-free flour is more likely to dry out. Such high temperatures can create patches of fully set, dry crumb on top of the muffins within the first 5 minutes of baking, leading to an uneven, skewed rise.

Cupcakes bake at a lower oven temperature of 160°C – for all the opposite reasons. Here, we want the edges and the centre to bake at rates as similar as possible, to create an even lift and a flat top.

FOR PERFECT CUPCAKES

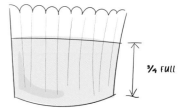

¾ FULL

FOR PERFECT MUFFINS

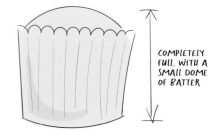

COMPLETELY FULL, WITH A SMALL DOME OF BATTER

Filling liners for cupcakes (top) and muffins (bottom) is not just a matter of pouring in batter any old how. Use this guide to achieve flat, only gently rounded cupcakes and beautifully domed muffins.

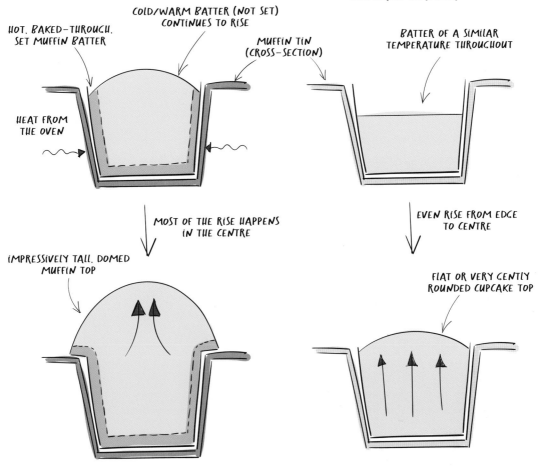

HIGH OVEN TEMPERATURE & COLD MUFFIN BATTER

HOT, BAKED-THROUGH, SET MUFFIN BATTER

COLD/WARM BATTER (NOT SET) CONTINUES TO RISE

MUFFIN TIN (CROSS-SECTION)

HEAT FROM THE OVEN

MOST OF THE RISE HAPPENS IN THE CENTRE

IMPRESSIVELY TALL, DOMED MUFFIN TOP

LOW OVEN TEMPERATURE & ROOM-TEMPERATURE CUPCAKE BATTER

BATTER OF A SIMILAR TEMPERATURE THROUGHOUT

EVEN RISE FROM EDGE TO CENTRE

FLAT OR VERY GENTLY ROUNDED CUPCAKE TOP

Oven temperature, ingredient temperature and the colour of the muffin tin all affect the rise and consequent shape of muffins (left) and cupcakes (right).

4. Ingredient temperature. Using room-temperature ingredients often seems to be one of the pillars of successful baking – and for cupcakes, it's 100% correct. Your eggs, butter and milk should all be at room temperature before you start. Muffins, however, are a different story. For these, it's best to use cold eggs, milk and yoghurt (you should still use room-temperature-softened butter, otherwise you'll have a difficult time working it into the flour with the reverse creaming method).

The reason for this difference is very closely related to point #3 about oven temperature. A reminder: if the edges set and fully bake through before the centre is done, the latter keeps on rising and we get a domed top. If you use cold ingredients, the centre of the muffin takes even longer to heat through, emphasising and exaggerating the effect of a higher oven temperature. Conversely, room-temperature ingredients mean that the batter temperature is much closer to the oven temperature, and the cupcake edges and centre heat up and bake at a more similar rate – helping you get a flat top, so working in the same direction as the lower oven temperature.

5. Bonus points: colour of the baking tin. This effect isn't quite as pronounced, and I definitely don't want you to go out and buy a different muffin tin if you already have a perfectly good one. In short, though, darker tins absorb and emit more heat than light ones. In the context of cupcakes and muffins, dark tins work a bit like a hotter 190°C oven and light tins like a cooler 160°C oven. Accordingly, for the best (of the best) results, use dark tins for muffins and light ones for cupcakes.

That said, if you follow the first four points, the colour of the baking tin doesn't have that much of an effect, and you'll still be able to bake beautiful, delicious cupcakes and muffins in either. I prefer a light baking tin because it gives a slower rate of caramelisation around the sides of the cupcakes than its darker counterparts. Even at 160°C, a dark baking tin can result in quite pronounced caramelisation at the sides and edges. I want to avoid this for things like vanilla or white chocolate cupcakes that look best if only slightly golden.

When it comes to muffins, a light tin also works, as the higher oven temperature is enough to produce a gorgeous golden-brown caramelisation (and, of course, a domed top) regardless of tin colour.

You'll notice that my top-five list doesn't refer to different amounts of raising agent nor to the presence of acidic ingredients, such as yoghurt, soured cream or lemon juice. That's because compared to things like batter consistency and oven temperature, these factors have a much less pronounced effect on the appearance and texture of the final bakes. Sure, muffins contain slightly more raising agents and, yes, they do contain more acidic ingredients, which can give the raising agents a boost, but the actual effect on the presence or absence of the domed top is debatable. The lemon poppy seed cupcakes on page 104, for example, aren't domed – and yet, we've added lemon juice to the batter.

The presence of yoghurt in muffin batter deserves a mention only insofar as it replaces part of the butter in cupcake recipes, creating a

Table 4: Cupcakes vs Muffins

	CUPCAKES	MUFFINS
batter consistency	loose, runny	thick
filling the liners	¾ full	to the brim
oven temperature	low, 160°C	high, 190°C
ingredient temperature	room temperature	cold
bonus points: colour of baking tin	light	dark

crumb that's less delicate and more bready. What's more, yoghurt (unlike milk) allows muffin batter to be thicker while ensuring the baked muffins stay nice and moist. Here's the thing: making gluten-free muffins is something of a baking balancing act – on the one hand, you want a fairly thick batter to get that beautiful bakery-style domed top. On the other, reducing the wet ingredients is always something of a hazard in gluten-free baking, in this case threatening dry, crumbly muffins. Using yoghurt or soured cream is therefore the perfect solution that, so to speak, combines the best of both worlds.

HOW TO PREVENT STICKING

If you haven't figured it out yet (you know, from my insistence on using plain gluten-free flour blends without xanthan added, unsalted butter, an oven thermometer and a digital scale – with decimal places), I like to be 1,000% in charge of how and why things happen. Leaving things to chance can be fun in the initial stages of recipe development when you're going with the flow to see what happens. But when you're following recipes with an expected outcome, precision is key.

That's why I find it hard to trust paper liners that claim to be non-stick. In my experience, the more confident the claim, the greater the likelihood that I'll sacrifice half the muffin to the paper-liner gods. My solution? Giving each liner a quick brush with sunflower or rapeseed oil. (Non-stick baking spray works, too.) Lightly coat the inside of the paper liner – just a gentle brush of oil, we don't want puddles – and stuck cupcakes and muffins will be a thing of the past.

Does it require an extra 2–3 minutes before you start baking? Sure. Is it a bit tedious? Definitely, but so is picking precious bits of cupcake off the liner. Does it save you tears of frustration? You bet.

A NOTE ON BUTTERCREAM

I've outlined my favourite buttercreams in the cake chapter (see page 44), and the same things hold true when it comes to icing cupcakes. What I do need to say, though, relates to the buttercream:cupcake ratio.

While my cupcakes don't exactly have a 1:1 buttercream:cupcake ratio, they certainly approach it. This might seem excessive, but remember that my buttercreams tend to be less sweet than the norm. And, honestly, it just looks pretty, especially when you finish it off with a sprinkling of chocolate shavings or chopped nuts.

The buttercream quantities in the recipes in this chapter give a buttercream:cupcake ratio as shown in the accompanying photos – if that seems a bit too much for your taste, just decrease all the buttercream ingredient amounts by the same factor. In the world of icing, there are really few firm rules and anything else is only a gentle guideline or suggestion, which you can tweak as you see fit.

THE IDEAL
CUPCAKE:BUTTERCREAM
RATIO
(NOTE: SUBJECTIVE. NOT SCIENTIFICALLY PROVEN)

1
:
1

TRIPLE CHOCOLATE CUPCAKES

MAKES 12
PREP TIME 1 hour 15 mins
BAKE TIME 26 mins
COOK TIME 5 mins

CHOCOLATE CUPCAKES

200g dark chocolate (60–70%
   cocoa solids), chopped
150g unsalted butter
15g Dutch processed cocoa
   powder
150g caster sugar
3 eggs, at room temperature
65g gluten-free flour blend
30g almond flour
1½ teaspoons baking powder
½ teaspoon xanthan gum
½ teaspoon salt
90g hot water ①
chocolate sprinkles or shavings,
   to decorate

CHOCOLATE FUDGE
FILLING

75g whole milk ②
75g dark chocolate (60–70% cocoa
   solids), chopped
40g caster sugar
25g Dutch processed cocoa
   powder
30g unsalted butter

CHOCOLATE
BUTTERCREAM

450g unsalted butter, softened
320g icing sugar
80g Dutch processed cocoa
   powder
½ teaspoon salt
200g dark chocolate (60–70%
   cocoa solids), melted and cooled

Don't let the large amounts of chocolate and surprisingly small amounts of flour alarm you. We're still in the Realm of Cupcakes, even if the ingredients' list might suggest we've crossed over into Brownie World. That said, if ever there were cupcakes that tasted like brownies (without actually being fudgy), with a moist crumb and an intense chocolate flavour, this is they. And because there's no such thing as too much chocolate (my personal motto), they're stuffed with a chocolate fudge filling, topped with chocolate buttercream and finished off with a sprinkling of chocolate shavings. I may call them 'triple' chocolate, but, honestly, I've lost count.

CHOCOLATE CUPCAKES

+ Adjust the oven shelf to the middle position, preheat the oven to 160°C and line a 12-hole muffin tin with cupcake liners.
+ In a heatproof bowl over a pan of simmering water, melt the dark chocolate and butter together. Set aside to cool until warm, then mix in the cocoa powder and sugar.
+ Add the eggs, one at a time, whisking well after each addition.
+ Sift the gluten-free flour blend, almond flour, baking powder, xanthan gum and salt into the chocolate mixture, and mix well until no flour clumps remain.
+ Add the hot water and whisk well until you get a smooth, glossy batter.
+ Divide the batter equally between the 12 cupcake liners, filling each about ¾ full.
+ Bake for 26–28 minutes or until an inserted toothpick comes out clean or with a few moist crumbs attached.
+ Remove the cupcakes from the tin and transfer to a wire rack to cool.

CHOCOLATE FUDGE FILLING

+ In a saucepan, cook the milk, half of the chopped chocolate, sugar and cocoa powder over a medium–high heat, stirring frequently until the chocolate has melted (about 5 minutes).
+ Remove from the heat and add the butter and the remaining chocolate. Mix until fully melted, then set aside until needed. The fudge sauce will thicken slightly while cooling.

CHOCOLATE BUTTERCREAM

+ In a stand mixer with the paddle attachment or using a hand mixer fitted with the double beaters, beat the butter for 2–3 minutes. Sift in the icing sugar and beat for a further 5 minutes until pale and fluffy.
+ Sift in the cocoa powder and salt, and beat until evenly distributed.
+ Add in the melted and cooled chocolate, and beat until you get a rich, fluffy chocolate buttercream that is an even, deep-chocolate brown.

ASSEMBLING THE CUPCAKES

+ Use an apple corer or the wide end of a piping nozzle to create a hole in the middle of each cupcake. Fill each with chocolate fudge sauce (if the sauce has thickened too much, microwave it in 10-second bursts until pourable).
+ Transfer the buttercream to a piping bag (use your choice of piping nozzle) and pipe on top of each cupcake, then sprinkle with chocolate sprinkles or shavings.

STORAGE

3–4 days in a closed container in a cool, dry place.

PUT A TWIST ON IT

You can pair the chocolate cupcakes with numerous alternative toppings, such as vanilla Swiss meringue buttercream (see page 92) or strawberry buttercream (see page 108). Or, try making:

S'mores cupcakes – Use crushed digestives (see page 162) mixed through with melted butter to create a base in each cupcake case. Pour the chocolate cupcake batter on top, then bake. Top with Swiss meringue, then toast it with a kitchen blow torch.

NOTES

① *The hot water gives an extra-moist, delicate crumb without making the finished cupcakes too wet or fudgy. And if you were worried about it scrambling or cooking the eggs, don't be. The eggs have been incorporated into the batter in a previous step and the amount of hot water is relatively low, so no scrambling will occur.*

② *You can add more milk depending on your desired filling consistency. 75g milk gives a thick, sauce-like filling that slowly firms up to fudginess.*

PHOTOGRAPHY OVERLEAF ⟶

VANILLA CUPCAKES

MAKES 12
PREP TIME 1 hour
BAKE TIME 22 mins

This recipe is the basis from which most other cupcakes follow. With minor tweaks (mostly to do with the wet:dry ingredients' ratio depending on the form of the flavourings you add), you can transform it into hundreds of other recipes. Still, I do rather love it as it is. Buttery, with a gentle vanilla flavour, of a gorgeous light golden colour and with a very gently domed top, this is my definition of what a cupcake ought to be.

VANILLA CUPCAKES

200g gluten-free flour blend

60g almond flour

200g caster sugar

3 teaspoons baking powder

½ teaspoon xanthan gum

¼ teaspoon salt

130g unsalted butter, softened

160g whole milk, at room temperature

2 eggs, at room temperature

2 egg whites, at room temperature ①

1 tablespoon lemon juice (optional) ②

1 teaspoon vanilla bean paste

sprinkles of choice, to decorate (optional)

VANILLA SWISS MERINGUE BUTTERCREAM

6 egg whites

375g caster sugar

¼ teaspoon cream of tartar

400g unsalted butter, softened

1 teaspoon vanilla bean paste

½ teaspoon salt

VANILLA CUPCAKES

+ Adjust the oven shelf to the middle position, preheat the oven to 160°C and line a 12-hole muffin tin with cupcake liners.

+ Sift together the gluten-free flour blend, almond flour, sugar, baking powder, xanthan gum and salt.

+ Add the butter and, using a stand mixer with the paddle attachment or a hand mixer fitted with the double beaters, work the butter into the dry ingredients until the texture resembles breadcrumbs. ③

+ In a separate bowl, whisk together the milk, eggs, egg whites, lemon juice (if using) and vanilla paste. Add the wet ingredients to the flour mixture in two batches, mixing well after each addition, until you get a smooth batter with no flour clumps.

+ Divide the batter equally between the 12 cupcake liners, filling each about ¾ full.

+ Bake for about 22–24 minutes or until an inserted toothpick comes out clean or with a few moist crumbs attached.

+ Remove the cupcakes from the muffin tin and transfer to a wire rack to cool.

VANILLA SWISS MERINGUE BUTTERCREAM

+ Mix the egg whites, sugar and cream of tartar in a heatproof bowl over a pan of simmering water. Stir continuously until the meringue mixture reaches 65°C and the sugar has dissolved.
+ Remove the mixture from the heat, transfer it to a stand mixer with a whisk attachment, or use a hand mixer fitted with the double beaters, and whisk on medium–high speed for 5–7 minutes until greatly increased in volume and at room temperature. It's very important that the meringue is at room temperature before you move on to the next step.
+ Add the butter, 1–2 tablespoons at a time, while mixing continuously on medium speed. If using a stand mixer, I recommend switching to the paddle attachment for this step. ④
+ Continue until you've used up all the butter and the buttercream looks smooth and fluffy. Add the vanilla paste and salt, mixing briefly to incorporate them.
+ Transfer the buttercream to a piping bag (use your choice of piping nozzle) and pipe on top of each cupcake, then decorate with sprinkles, if you wish.

STORAGE

3–4 days in a closed container in a cool, dry place.

PUT A TWIST ON IT

Pair the vanilla cupcakes with numerous other toppings, such as chocolate buttercream (see page 88) or strawberry buttercream (see page 108). This basic recipe was also the foundation for the lemon poppy seed cupcakes (see page 104), strawberry and white chocolate cupcakes (see page 108) and coffee coffee coffee cupcakes (see page 98).

NOTES

① *The extra egg whites make the cupcakes especially fluffy without adding more raising agents, which could negatively impact the flavour.*

② *Adding a teaspoon of lemon juice is optional, but its acidity gives an extra boost to the raising agents, resulting in even fluffier cupcakes without impacting the vanilla flavour. Although some recipes use yoghurt or soured cream to do the same job, those additions have a greater effect on flavour.*

③ *The reverse creaming method (see pages 38–9) gives the cupcakes a smooth, velvety crumb and an even aeration, while helping achieve a flat top.*

④ *Add the butter slowly, so that the amount of fat doesn't overwhelm the meringue. You're creating an emulsion, which is fairly delicate and all about balance and timing. If the buttercream goes through a 'soupy' stage, where it's very runny and looks split, continue adding butter and whisking/ beating the mixture. It will come together once you've added all the butter. For tips on troubleshooting Swiss meringue buttercream, see page 44.*

PHOTOGRAPHY OVERLEAF ⟶

PEANUT BUTTER CHOCOLATE CUPCAKES

MAKES 14
PREP TIME 45 mins
BAKE TIME 24 mins

PEANUT BUTTER CUPCAKES

160g gluten-free flour blend
40g almond flour
200g light brown soft sugar
3 teaspoons baking powder
½ teaspoon xanthan gum
½ teaspoon salt
150g natural, unsweetened,
    smooth peanut butter
130g sunflower or other
    neutral-tasting oil ①
160g whole milk, at room
    temperature
2 eggs, at room temperature
2 egg whites, at room temperature
1 tablespoon lemon juice
15g raw, unsalted peanuts, toasted
    and chopped, to decorate

PEANUT BUTTER CHOCOLATE BUTTERCREAM

350g unsalted butter, softened
300g natural, unsweetened,
    smooth peanut butter
300g icing sugar
150g dark chocolate (60–70%
    cocoa solids), melted and cooled
45g Dutch processed cocoa powder
½ teaspoon salt

NOTE

① *Adjusting oil is crucial for flat*
   peanut-butter cupcakes. Without
   it, the final batter is thicker and
   behaves more like muffin batter.

There's just something about the combination of chocolate and peanut butter that feels terribly decadent. Both these flavours are rich and intense on their own, but together in these cupcakes they create something even more delicious. The cupcakes illustrate how using peanut butter as the only fat component in baking is usually rather tricky. To achieve perfectly delicate, fluffy results, you need a mixture of peanut butter and oil. Similarly, adding butter to the peanut butter chocolate buttercream gives a much airier texture than just peanut butter alone. Still, these additional ingredients in no way mask the peanut-butter flavour. Rather, they allow it to shine.

PEANUT BUTTER CUPCAKES

+   Adjust the oven shelf to the middle position, preheat the oven to 160°C and line a 12-hole muffin tin with cupcake liners. Line a second muffin tin with 2 liners (this recipe makes 14 cupcakes).
+   Sift together the gluten-free flour blend, almond flour, sugar, baking powder, xanthan gum and salt.
+   In a separate bowl, whisk together the peanut butter, oil, milk, eggs, egg whites and lemon juice. Add the wet ingredients to the flour mixture and whisk until you get a smooth batter with no flour clumps.
+   Divide the batter equally between the 14 cupcake liners, filling each about ¾ full.
+   Bake for about 24–26 minutes or until an inserted toothpick comes out clean or with a few moist crumbs attached.
+   Remove the cupcakes from the tin and transfer to a wire rack to cool.

PEANUT BUTTER CHOCOLATE BUTTERCREAM

+   Using a stand mixer with the paddle attachment or a hand mixer fitted with the double beaters, beat the butter and peanut butter on medium speed for 2–3 minutes until evenly combined and fluffy.
+   Sift in the icing sugar and beat for a further 5–7 minutes until pale in colour and very fluffy, then add the chocolate, cocoa powder and salt, and mix until incorporated (about 1–2 minutes).
+   Transfer the buttercream to a piping bag (use your choice of piping nozzle) and pipe on top of each cupcake, then sprinkle with the chopped toasted peanuts to finish.

STORAGE

3–4 days in a closed container in a cool, dry place.

PUT A TWIST ON IT

Peanut butter + jam cupcakes – Pair the peanut butter cupcakes with a strawberry-jam centre and strawberry buttercream (see page 108). Or, try raspberry jam and add freeze-dried raspberries to the buttercream.

COFFEE COFFEE COFFEE CUPCAKES

MAKES 12
PREP TIME 1 hour 15 mins
BAKE TIME 22 mins
COOK TIME 10 mins

COFFEE CUPCAKES

160g whole milk
1½–2 tablespoons instant coffee
    granules ①
200g gluten-free flour blend
60g almond flour
200g caster sugar
3 teaspoons baking powder
½ teaspoon xanthan gum
¼ teaspoon salt
130g unsalted butter, softened
2 eggs, at room temperature
2 egg whites, at room temperature

COFFEE CARAMEL SAUCE

90g double cream
½–1 tablespoon instant coffee
    granules
100g caster sugar
35g unsalted butter
¼ teaspoon salt (optional)

COFFEE FROSTING

550g double cream, chilled
4½ tablespoons cappuccino
    powder
2½ teaspoons instant coffee
    granules
300g icing sugar, sifted

Coffee lovers, rejoice. These cupcakes just might be the ultimate coffee dessert, with three different coffee components that come together in a coffee taste explosion. You might be tempted to skip the cappuccino powder in the frosting – don't. It's the cappuccino that takes the topping to a whole new level, by smoothing out the more robust, intense notes of instant coffee. As for the coffee caramel sauce: it's a revelation and once you have a taste, you'll be tempted to add it to… well, everything.

COFFEE CUPCAKES

+   Adjust the oven shelf to the middle position, preheat the oven to 160°C and line a 12-hole muffin tin with cupcake liners.
+   Heat the milk and instant coffee granules together until the coffee is fully dissolved. Set aside to cool until warm or room temperature.
+   Sift together the gluten-free flour blend, almond flour, caster sugar, baking powder, xanthan gum and salt.
+   Add the butter and, using a stand mixer with the paddle attachment or a hand mixer fitted with the double beaters, work the butter into the dry ingredients until the texture resembles breadcrumbs.
+   In a separate bowl, whisk together the coffee-infused milk, eggs and egg whites. Add the wet ingredients to the flour mixture in two batches, mixing well after each addition, until you get a smooth batter with no flour clumps.
+   Divide the batter equally between the 12 cupcake liners, filling each about ¾ full.
+   Bake for about 22–24 minutes or until an inserted toothpick comes out clean or with a few moist crumbs attached.
+   Remove the cupcakes from the tin and transfer to a wire rack to cool.

COFFEE CARAMEL SAUCE

+   In a small saucepan or microwave-safe bowl, heat the double cream with the instant coffee granules until dissolved. ② Set aside until needed.
+   In a separate saucepan, cook the sugar over a medium heat, stirring occasionally, until melted and caramelised to a light amber colour. ③
+   Add the butter and mix well until the butter has completely melted, then cook for a further 30 seconds.
+   Remove from the heat and pour in the coffee cream, mixing well to incorporate. Mix in the salt (if using).
+   Cool to room temperature, so that the caramel sauce thickens slightly.

COFFEE FROSTING

+ In a small saucepan or a microwave-safe bowl, heat together 110g of the double cream with the cappuccino powder and instant coffee granules until dissolved. Set aside to cool completely.
+ Using a stand mixer with the whisk attachment or a hand mixer fitted with the double beaters, whip the remaining double cream and the icing sugar until stiff peaks form.
+ Add the coffee cream, 1 tablespoon at a time, whisking continuously on medium speed. Once you have added all the coffee cream, whip for a further 30 seconds until completely incorporated and the frosting holds stiff peaks. ④

ASSEMBLING THE CUPCAKES

+ Use an apple corer or the large end of a piping nozzle to create a hole in the middle of each cupcake. Fill each with coffee caramel sauce (if the sauce has thickened too much, microwave it in 10-second busts until pourable or spoonable).
+ Transfer the coffee frosting to a piping bag (use your choice of piping nozzle) and pipe on top of each cupcake.

STORAGE

3–4 days in a closed container in a cool, dry place.

NOTES

① *As well as adding flavour, coffee is on the acidic end of the pH spectrum so gives a boost to the raising agents for extra-fluffy and delicate cupcakes.*

② *Adding the coffee to the double cream (rather than to the sugar before caramelising) is important because it means you can caramelise the sugar on its own and in that way keep an eye on its colour – you need to be able to tell when the caramel turns the right shade of amber without the darkness of the coffee getting in the way.*

③ *A caramel that's light in colour and so also in flavour works well with the rich, slightly bitter coffee notes.*

④ *Preparing the frosting in this order is important to keep it stable – it makes it easy to pipe and the result holds its shape well. If you tried to whip the coffee cream together with the rest of the cream and sugar from the very beginning, you may end up with a soupy, split frosting.*

PHOTOGRAPHY OVERLEAF ⟶

CARROT CAKE CUPCAKES

MAKES 12
PREP TIME 45 mins
BAKE TIME 24 mins

CARROT CAKE CUPCAKES
160g gluten-free flour blend
160g light brown soft sugar
1 teaspoon baking powder
½ teaspoon bicarbonate of soda
½ teaspoon xanthan gum
1 teaspoon ground cinnamon
¼ teaspoon ground ginger
¼ teaspoon ground mixed spice
¼ teaspoon salt
130g sunflower or other
    neutral-tasting oil
60g full-fat plain or Greek yoghurt,
    at room temperature
2 eggs, at room temperature
200g carrots, coarsely grated
60g walnuts, chopped, plus extra
    to decorate

CREAM-CHEESE
FROSTING ①
300g double cream, chilled
150g icing sugar, sifted
300g full-fat cream cheese, chilled
1 teaspoon vanilla bean paste
¼ teaspoon salt

NOTE
① *Double cream makes a more*
   stable cream-cheese frosting than
   you could achieve using butter,
   even without large amounts of
   icing sugar (see page 45).

These cupcakes are beautifully moist and flavourful thanks to the grated carrots (coarsely grating is essential – finely grated carrots will release too much moisture during baking and result in dense cupcakes) and an abundance of spices. They have a wonderful added crunch from the chopped walnuts. The cream-cheese frosting pipes like a dream and tastes like a cloud of fluffy, sweet, cream-cheese goodness. A fun alternative to making a single carrot cake, these are the perfect addition to any party or afternoon tea.

CARROT CAKE CUPCAKES
+   Adjust the oven shelf to the middle position, preheat the oven to 160°C and line a 12-hole muffin tin with cupcake liners.
+   Sift together the gluten-free flour blend, sugar, baking powder, bicarbonate of soda, xanthan gum, spices and salt.
+   In a separate bowl, whisk together the oil, yoghurt and eggs. Add the wet ingredients to the flour mixture and whisk until you get a smooth, thick batter with no flour clumps.
+   Fold in the grated carrots and chopped walnuts until evenly distributed.
+   Divide the batter equally between the 12 cupcake liners, filling each about ⅔ full.
+   Bake for about 24–26 minutes or until an inserted toothpick comes out clean or with a few moist crumbs attached.
+   Remove the cupcakes from the tin and transfer to a wire rack to cool.

CREAM-CHEESE FROSTING
+   Using a stand mixer with the whisk attachment or a hand mixer fitted with the double beaters, whip the double cream and icing sugar until the mixture holds stiff peaks.
+   In a separate bowl, whip the cream cheese until smooth. Add it to the whipped cream and whip for about 1 minute until the mixture holds stiff peaks. Add the vanilla paste and salt, and mix until incorporated.
+   Transfer the frosting to a piping bag (use your choice of piping nozzle) and pipe on top of each cupcake, then sprinkle with the chopped walnuts to finish.

STORAGE
3–4 days in a closed container in a cool, dry place.

LEMON POPPY SEED CUPCAKES

MAKES 12
PREP TIME 1 hour 30 mins
BAKE TIME 22 mins
COOK TIME 6 mins

I like to describe these cupcakes as sunshine in dessert form. They make people happy, simple as that. From the very first bite, when you encounter the fluffiness of the Swiss meringue buttercream, to the burst of lemon zest in the sponge, there's something special about them. And then, of course, there's the surprise – the detail that makes them out of this world: a centre of bright, refreshing lemon curd.

LEMON POPPY SEED CUPCAKES

200g gluten-free flour blend
60g almond flour
200g caster sugar
3 teaspoons baking powder
½ teaspoon xanthan gum
¼ teaspoon salt
4 tablespoons poppy seeds
zest of 2 unwaxed lemons
130g unsalted butter, softened
160g whole milk, at room
    temperature
2 eggs, at room temperature
2 egg whites, at room
    temperature ①
2 tablespoons lemon juice
½ teaspoon vanilla bean paste

LEMON CURD

60g lemon juice
75g caster sugar
¼ teaspoon salt
3 egg yolks
50g unsalted butter

POPPY SEED SWISS MERINGUE BUTTERCREAM

6 egg whites
375g caster sugar
¼ teaspoon cream of tartar
400g unsalted butter, softened
½ teaspoon salt
2–4 tablespoons poppy seeds

LEMON POPPY SEED CUPCAKES

+ Adjust the oven shelf to the middle position, preheat the oven to 160°C and line a 12-hole muffin tin with cupcake liners.
+ Sift together the gluten-free flour blend, almond flour, caster sugar, baking powder, xanthan gum and salt. Mix in the poppy seeds and lemon zest.
+ Add the butter and, using a stand mixer with the paddle attachment or a hand mixer fitted with the double beaters, work the butter into the dry ingredients until the texture resembles breadcrumbs.
+ In a separate bowl, whisk together the milk, eggs, egg whites, lemon juice and vanilla paste. Add the wet ingredients to the flour mixture in two batches, mixing well after each addition, until you get a smooth batter with no flour clumps.
+ Divide the batter equally between the 12 cupcake liners, filling each about ¾ full.
+ Bake for 22–24 minutes or until an inserted toothpick comes out clean or with a few moist crumbs attached.
+ Remove the cupcakes from the tin and transfer to a wire rack to cool.

LEMON CURD

+ Combine all the lemon curd ingredients except the butter in a saucepan and whisk briefly until the egg yolks are evenly incorporated. Cook over a medium–high heat, stirring (not whisking) frequently, for about 6–8 minutes until the lemon curd starts thickening. (The temperature should be at about 76–78°C on a cooking thermometer.) ②
+ Remove from the heat and stir in the butter, mixing well until melted. (Alternatively, for extra-smooth lemon curd, you can pass the mixture through a sieve before adding the butter.)
+ Allow to cool, stirring occasionally to prevent a skin forming.

POPPY SEED SWISS MERINGUE BUTTERCREAM

+ Mix the egg whites, sugar and cream of tartar in a heatproof bowl over a pan of simmering water. Stir continuously until the meringue mixture reaches 65°C and the sugar has dissolved.
+ Remove the mixture from the heat, transfer it to a stand mixer with a whisk attachment, or use a hand mixer fitted with the double beaters, and whisk on medium–high speed for 5–7 minutes until greatly increased in volume and at room temperature. It's very important that the meringue is at room temperature before you move on to the next step.
+ Add the butter, 1–2 tablespoons at a time, while mixing continuously on medium speed. If using a stand mixer, I recommend switching to the paddle attachment for this step. ③
+ Continue until you've used up all the butter and the buttercream looks smooth and fluffy. Add the salt and poppy seeds, mixing briefly to incorporate them.

ASSEMBLING THE CUPCAKES

+ Use an apple corer or the large end of a piping nozzle to create a hole in the middle of each cupcake. Fill each with cooled lemon curd.
+ Transfer the buttercream to a piping bag (use your choice of piping nozzle) and pipe on top of each cupcake.

STORAGE

3–4 days in a closed container in a cool, dry place.

NOTES

① *The extra egg whites make the cupcakes especially fluffy without adding more raising agents, which could negatively impact the flavour.*

② *Stirring the curd using a wooden spoon or a spatula (rather than whisking it) helps prevent the formation of foam, which is pretty impossible to get rid of, even when the curd starts thickening.*

③ *Using the paddle attachment, rather than the whisk, incorporates less air into the buttercream at this point to give a smooth, velvety texture. For tips on troubleshooting Swiss meringue buttercream, see page 44.*

PHOTOGRAPHY OVERLEAF ⟶

STRAWBERRY + WHITE CHOCOLATE CUPCAKES

MAKES 16
PREP TIME 1 hour
BAKE TIME 22 mins

The slight tartness of strawberries here balances the rich sweetness of white chocolate. The cupcakes are buttery and fluffy; the strawberry-jam centre is a lovely surprise. What really makes them memorable, though, is the topping. The trick to an amazing strawberry buttercream (without any artificial flavourings) lies in using freeze-dried strawberries, blitzed with icing sugar to a powder.

WHITE CHOCOLATE CUPCAKES

200g gluten-free flour blend
60g almond flour
150g caster sugar
3 teaspoons baking powder
½ teaspoon xanthan gum
¼ teaspoon salt
60g unsalted butter, softened
160g whole milk, at room temperature
2 eggs, at room temperature
2 egg whites, at room temperature
150g white chocolate, melted and cooled until warm ①
8–16 tablespoons strawberry jam, to fill
white chocolate shavings or sprinkles, to decorate

STRAWBERRY BUTTERCREAM

55g freeze-dried strawberries
530g icing sugar
620g unsalted butter, softened
1–2 drops of red food colouring (optional)

NOTE

① *Melting the white chocolate separately preserves its off-white colour and prevents it splitting, as it might if melted with butter. Compared to dark chocolate, white chocolate is more sensitive to the moisture in butter.*

WHITE CHOCOLATE CUPCAKES

+ Adjust the oven shelf to the middle position, preheat the oven to 160°C and line a 12-hole muffin tin with cupcake liners. Line a second muffin tin with 4 liners (this recipe makes 16 cupcakes).
+ Sift together the gluten-free flour blend, almond flour, caster sugar, baking powder, xanthan gum and salt. Add the butter and, using a stand mixer with the paddle attachment or a hand mixer fitted with the double beaters, work the butter into the dry ingredients until the texture resembles breadcrumbs.
+ In a separate bowl, whisk together the milk, eggs and egg whites. Add the wet ingredients to the flour mixture and whisk until you get a smooth batter with no flour clumps.
+ Add the white chocolate and mix until fully incorporated. Divide the batter equally between the 16 cupcake liners, filling each about ⅔ full.
+ Bake for about 22–24 minutes or until an inserted toothpick comes out clean or with a few moist crumbs attached.
+ Remove the cupcakes from the tin and transfer to a wire rack to cool.

STRAWBERRY BUTTERCREAM

+ In a food processor or blender, blitz the freeze-dried strawberries and 30g of the icing sugar until fine. Pass the powder through a sieve to remove any larger pieces, then set aside until needed.
+ Using a stand mixer with the paddle attachment or a hand mixer fitted with the double beaters, beat the butter for 2–3 minutes. Sift in the remaining icing sugar along with the strawberry sugar, and beat for a further 5 minutes until fluffy and pink.
+ Add the red food colouring for a more intense colour, if you wish.

ASSEMBLING THE CUPCAKES

+ Use an apple corer or the large end of a piping nozzle to create a hole in the middle of each cupcake. Fill each with ½–1 tablespoon of jam.
+ Transfer the buttercream to a piping bag (use your choice of nozzle) and pipe on to each cupcake. Decorate with the shavings or sprinkles.

STORAGE

3–4 days in a closed container in a cool, dry place.

DOUBLE CHOCOLATE MUFFINS

MAKES 12
PREP TIME 30 mins
BAKE TIME 20 mins

Double the chocolate, double the fun – that's pretty much my life's motto, and these muffins are all the more delicious for it. They are moist, with an intense chocolate flavour thanks to both a generous amount of melted chocolate and a few spoonfuls of Dutch processed cocoa powder. The chocolate chips complete what is an explosion of irresistible chocolatey goodness.

150g dark chocolate (60–70% cocoa solids), chopped ①
125g unsalted butter, softened
180g gluten-free flour blend
60g almond flour
30g Dutch processed cocoa powder
200g caster sugar
2 teaspoons baking powder
1 teaspoon bicarbonate of soda
1 teaspoon xanthan gum
½ teaspoon salt
100g whole milk
100g full-fat plain or Greek yoghurt
2 eggs
150g dark or milk chocolate chips

+ Adjust the oven shelf to the middle position, preheat the oven to 190°C and line a 12-hole muffin tin with muffin liners.
+ In a heatproof bowl over a pan of simmering water, melt the dark chocolate and butter together. Set aside to cool until warm.
+ Sift together the gluten-free flour blend, almond flour, cocoa powder, sugar, baking powder, bicarbonate of soda, xanthan gum and salt.
+ In a separate bowl, whisk together the milk, yoghurt and eggs. Add the wet ingredients and the melted chocolate mixture to the dry ingredients, and whisk until you get a smooth, thick batter with no flour clumps.
+ Reserve some of the chocolate chips to sprinkle on top and fold in the remainder. ②
+ Divide the batter equally between the 12 muffin liners, filling each to the brim.
+ Sprinkle the reserved chocolate chips on top of the muffins, and bake for about 20–22 minutes or until an inserted toothpick comes out clean or with a few moist crumbs attached.
+ Remove the muffins from the tin and transfer to a wire rack to cool.

NOTES

① *Because chocolate is slightly acidic, it reacts with the raising agents in the muffin batter giving even taller and more domed muffins than their vanilla counterpart.*

② *Because the muffin batter is fairly thick, there's no need to worry about the chocolate chips sinking – they will remain evenly distributed throughout baking.*

STORAGE

Best on the day, but you can keep them in a closed container in a cool, dry place for up to 4 days. If you're serving on days 3 or 4, reheat in the microwave for 5–10 seconds first.

BLUEBERRY MUFFINS

MAKES 12
PREP TIME 30 mins
BAKE TIME 20 mins

280g gluten-free flour blend
80g almond flour
150g light brown soft sugar
2 teaspoons baking powder
1 teaspoon bicarbonate of soda
1 teaspoon xanthan gum
½ teaspoon salt
zest of 2 unwaxed lemons
130g unsalted butter, softened
140g whole milk, cold ①
140g full-fat plain or Greek
    yoghurt, cold
2 eggs, cold
2 tablespoons lemon juice
1 teaspoon vanilla bean paste
180g fresh or frozen blueberries
1 tablespoon granulated sugar,
    for sprinkling

NOTES

① *Cold ingredients (including
    the yoghurt and eggs) help
    with the formation of the
    dramatically domed muffin
    top (see pages 85–6).*

② *Muffin batter is significantly
    thicker than cupcake batter,
    which helps muffins rise and
    form the characteristic domed
    muffin top (see page 84).*

③ *This ensures that even if some
    blueberries do sink slightly
    during baking, the distribution
    of the fruit is fairly even
    throughout the muffins.*

A perfect blueberry muffin has to be quick to whip up, have a
beautifully domed, caramelised top, be absolutely loaded with
blueberries, and have a moist, tender crumb. The gluten-free blueberry
muffins you'll get from this recipe tick all those boxes. An extra
sprinkling of granulated sugar just before they go into the oven
gives them a pretty sparkle, and a sweet crunch as you bite into them.

+   Adjust the oven shelf to the middle position, preheat the oven to 190°C
    and line a 12-hole muffin tin with muffin liners.
+   Sift together the gluten-free flour blend, almond flour, brown sugar,
    baking powder, bicarbonate of soda, xanthan gum and salt.
+   Add the lemon zest and butter and, using a stand mixer with the paddle
    attachment or a hand mixer fitted with the double beaters, work the
    butter into the dry ingredients until the texture resembles breadcrumbs.
+   In a separate bowl, whisk together the milk, yoghurt, eggs, lemon juice
    and vanilla paste. Add the wet ingredients to the flour mixture and
    whisk until you get a smooth, thick batter with no flour clumps. ②
+   Spoon about 1 tablespoon of the blueberry-free cupcake batter into the
    bottom of each cupcake liner. ③
+   Set aside about 30g of the blueberries to sprinkle over the top of the
    muffins. Fold the remainder into the remaining batter in the bowl.
    Divide the batter equally between the 12 muffin liners, filling each
    to the brim.
+   Scatter the reserved blueberries on top of the muffins, sprinkle with
    granulated sugar, and bake for about 20–22 minutes (if using fresh
    blueberries) or 24–26 minutes (if using frozen), or until an inserted
    toothpick comes out clean or with a few moist crumbs attached.
+   Remove the muffins from the tin and transfer to a wire rack to cool.

STORAGE
Best on the day, but you can keep them in a closed container in a cool,
dry place for up to 4 days. If you're serving on days 3 or 4, reheat in the
microwave for 5–10 seconds first.

PUT A TWIST ON IT
Other berries, such as raspberries or blackberries work well in these
muffins, too. Or, try:
Chocolate chip muffins – Instead of the blueberries, simply add dark
or milk chocolate chips.

APPLE PIE MUFFINS

MAKES 12
PREP TIME 45 mins
CHILL TIME 30 mins
BAKE TIME 22 mins

CINNAMON CRUMBLE
70g gluten-free flour blend
50g light brown soft sugar
½ teaspoon ground cinnamon
40g unsalted butter, cold

APPLE MUFFINS
2 sweet–tart, firm apples (such
    as Pink Lady, Granny Smith
    or Braeburn)
2 tablespoons lemon juice
175g light brown soft sugar
1½ teaspoons ground cinnamon
280g gluten-free flour blend
80g almond flour
2 teaspoons baking powder
1 teaspoon bicarbonate of soda
1 teaspoon xanthan gum
½ teaspoon ground ginger
½ teaspoon ground nutmeg
½ teaspoon salt
130g unsalted butter, softened
140g whole milk, cold ①
140g full-fat plain or Greek
    yoghurt, cold
2 eggs, cold
100g pecans, chopped

NOTES
① *Cold ingredients (including the
yoghurt and eggs) help with the
formation of the domed muffin
top (see pages 85–6).*

② *Refrigerating the crumble before
use ensures that it keeps its form
during baking.*

While these muffins have the most wonderful, moist, cinnamon-y crumb, dotted with chopped apples, it's the crumble top that really makes them special. It's buttery and crunchy, and the chopped pecans get nicely toasted in the oven for an extra layer of flavour.

CINNAMON CRUMBLE
+ In a bowl, mix together the gluten-free flour blend, sugar and cinnamon. Add the butter and work it into the flour mixture until crumbly but it holds its shape when pressed together.
+ Refrigerate the crumble mixture for at least 30 minutes while you prepare the muffin batter. It should be firm or solid to the touch when you sprinkle it on top of the muffins. ②

APPLE MUFFINS
+ Adjust the oven shelf to the middle position, preheat the oven to 190°C and line a 12-hole muffin tin with muffin liners.
+ Peel and core the apples, then cube them to about the size of peas. Toss the pieces with the lemon juice, 25g of the sugar and 1 teaspoon of the cinnamon. Set aside to macerate and release any juices.
+ Sift together the gluten-free flour blend, almond flour and remaining sugar, along with the baking powder, bicarbonate of soda, xanthan gum, remaining cinnamon, and the ginger, nutmeg and salt.
+ Add the butter and, using a stand mixer with the paddle attachment or a hand mixer fitted with the double beaters, work the butter into the dry ingredients until the texture resembles breadcrumbs or coarse sand.
+ In a separate bowl, whisk together the milk, yoghurt and eggs. Add the wet ingredients to the flour mixture, and whisk until you get a smooth, thick batter with no flour clumps. Fold in the chopped apples along with any released juices and 75g of the chopped pecans. Divide the batter equally between the 12 muffin liners, filling each to the brim.
+ Mix the chilled crumble with the remaining pecans, and sprinkle it generously on top of the muffins. Make sure that the majority of the crumble is in a heap in the middle of the muffins with less around the edge, to account for the rise of the muffin top during baking.
+ Bake for about 22–24 minutes or until an inserted toothpick comes out clean or with a few moist crumbs attached.
+ Remove the muffins from the tin and transfer to a wire rack to cool.

STORAGE
Best on the day, but you can keep them in a closed container in a cool, dry place for up to 4 days. If you're serving on days 3 or 4, reheat in the microwave for 5–10 seconds first.

BROWN BUTTER BANANA NUT MUFFINS

MAKES 12
PREP TIME 30 mins
COOK TIME 6 mins
BAKE TIME 20 mins

2 bananas, mashed (about 200g,
    peeled weight)
2 tablespoons lemon juice
150g unsalted butter
150g full-fat plain or Greek yoghurt
100g whole milk
2 eggs
280g gluten-free flour blend
80g almond flour
150g light brown soft sugar
2 teaspoons baking powder
1 teaspoon bicarbonate of soda
1 teaspoon xanthan gum
¼ teaspoon salt
90g pecans, roughly chopped

I would like to give a big hug to whomever first thought to cook butter until it turns amber brown and nutty and so even more delicious than butter already is. The science behind browning butter is to do with the Maillard reaction (which is also what makes a deep golden-brown crust on bread so delicious and gives seared steaks their flavour) – but beyond the science, the nutty tones of browned butter pair wonderfully with the mashed bananas and the pecans in these moist, flavourful muffins. The bananas make the muffin tops extra delicious, giving the sugar a helping hand as it caramelises.

+ Adjust the oven shelf to the middle position, preheat the oven to 190°C and line a 12-hole muffin tin with muffin liners.
+ Mix the bananas with the lemon juice to prevent oxidation and browning, and set aside until needed.
+ To brown the butter, add it to a saucepan (preferably one of a light colour so you can see the butter changing colour) and cook it over a medium heat, stirring frequently, for about 5–6 minutes until melted. It will then start bubbling and foaming, and turn amber. You should see specks of a deep brown colour on the bottom of the saucepan – those are the caramelised milk solids. This should take about 6–8 minutes in total. Remove the butter from the heat and set aside to cool until warm.
+ Mix together the cooled browned butter with the mashed bananas, yoghurt, milk and eggs.
+ In a separate bowl, sift together the gluten-free flour blend, almond flour, sugar, baking powder, bicarbonate of soda, xanthan gum and salt.
+ Add the wet ingredients to the flour mixture and whisk to a smooth, thick batter with no flour clumps. Set aside some of the pecans for sprinkling, then mix the remainder into the batter.
+ Divide the batter equally between the 12 muffin liners, filling each to the brim.
+ Sprinkle the reserved pecans on top of the muffins and bake for about 20–22 minutes or until golden brown and an inserted toothpick comes out clean or with a few moist crumbs attached.
+ Remove the muffins from the tin and transfer to a wire rack to cool.

STORAGE
Best on the day, but you can keep them in a closed container in a cool, dry place for up to 4 days. If you're serving on days 3 or 4, reheat in the microwave for 5–10 seconds first.

PUT A TWIST ON IT
Choco-hazelnut banana nut muffins – Swap the pecans for hazelnuts and swirl a spoonful of chocolate hazelnut spread on top of each muffin before baking.

COURGETTE + FETA MUFFINS

MAKES 10
PREP TIME 20 mins
BAKE TIME 30 mins

120g unsalted butter, melted and
    cooled, plus extra for greasing
2–3 courgettes, coarsely grated
    (about 300g, prepared weight)
300g gluten-free flour blend ①
2 teaspoons baking powder
½ teaspoon xanthan gum
½–1 teaspoon salt
½ teaspoon black and white
    pepper
1 tablespoon thyme and rosemary
    leaves, chopped
4 eggs, at room temperature
60g whole milk, at room
    temperature
150g feta cheese, crumbled
120g sun-dried tomatoes, roughly
    chopped

NOTE

① *Courgettes comprise about
95% water. A relatively large
amount of gluten-free flour blend
makes the batter fairly thick
to prevent the muffins from
becoming soggy once the
courgettes release any further
moisture during baking. (The
courgettes' high water content
is why it is also important to
squeeze them out before adding
them to the batter.)*

The world of savoury muffins is as varied as it is delicious, but
these courgette and feta muffins are my absolute favourites. They're
beautifully moist thanks to the grated courgette, while the saltiness
of the feta cheese and the wonderful aroma of sun-dried tomatoes
take them to a whole new level. At the same time, they're incredibly
easy and quick to prepare, making them the perfect snack or even
a delicious addition to breakfast or brunch.

+ Adjust the oven shelf to the middle position, preheat the oven to
  190°C and generously butter 10 holes of a 12-hole muffin tin.
+ Place the grated courgettes into a sieve over a bowl and allow any
  excess juice to drip out – press down gently to squeeze out as much
  moisture as possible.
+ Sift together the gluten-free flour blend, baking powder, xanthan gum,
  salt and pepper. Add the herbs and mix well.
+ In a separate bowl, whisk together the melted butter, eggs and milk.
  Add the wet ingredients to the flour mixture and whisk until you get a
  smooth, thick batter with no flour clumps.
+ Add the drained grated courgettes, feta cheese and sun-dried tomatoes,
  and mix until well combined.
+ Divide the batter equally between the 10 muffin-tin holes, filling
  each about ¾ full. Bake for about 30–32 minutes or until golden brown
  and an inserted toothpick comes out clean or with a few moist crumbs
  attached.
+ Remove the muffins from the tin and transfer to a wire rack to cool.

STORAGE

Best on the day, but you can keep them in a closed container in a cool,
dry place for up to 4 days. If you're serving on days 3 or 4, reheat in the
microwave for 5–10 seconds first.

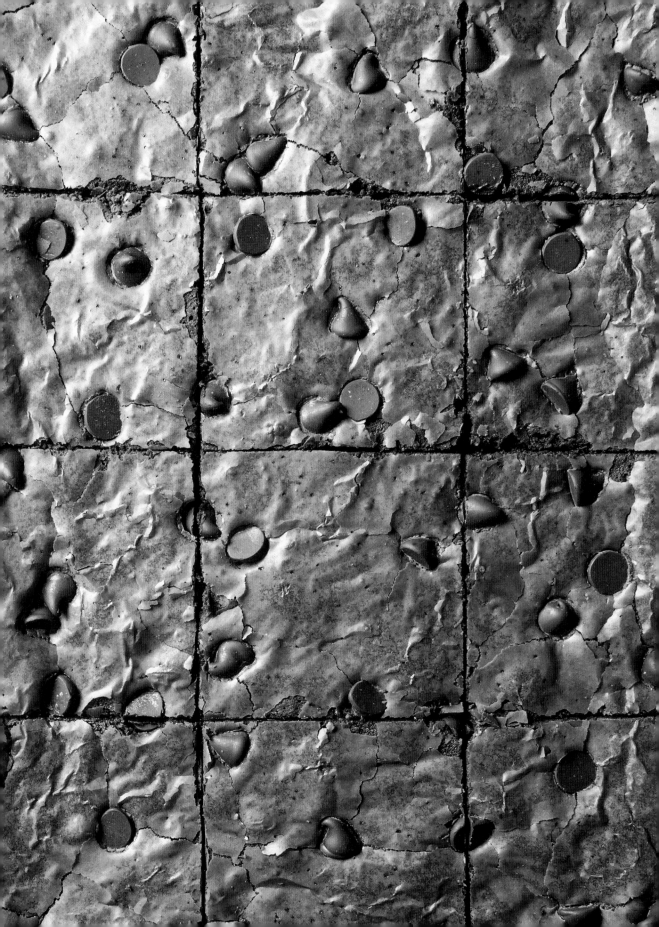

BROWNIES

Let me start by saying that I take brownies very seriously. As much as I adore cake and cookies, brownies are my one, true food love. There's just something about their perfectly dense, fudgy texture and that elusive, much-coveted, paper-thin shiny crust that sends chills down my spine.

And then, of course, there's the chocolate. Oh – the chocolate. I could write essays about the seductive glossiness of chocolate as it's melted with butter, or the magic that is chocolate being oh-so gently folded into whisked eggs and sugar… but don't worry, I'll try to restrain myself.

Suffice it to say I've devoted months of my life to finding the ideal formulation to get the perfect shiny top brownies (on the last count, I was on recipe variation #21) – and you get to enjoy the results of my brownie obsession. As always, you're welcome.

THE TYPES OF BROWNIE IN THIS CHAPTER

Considering my love of brownies, it's hardly surprising that instead of just one or two recipes, brownies get a chapter all of their own – with no cookie bars or other sweet treats getting in the way of chocolate (over)indulgence.

Overall, there are five basic brownie recipes in this chapter: shiny top, crackly top, cocoa, grain-free and white chocolate. These recipes, while dreamy just as they are, are easy to adapt further into numerous other recipes – all you have to do is add things such as spices, nuts or chocolate chips (because more chocolate is always a good idea).

Then, there are a few other brownie recipes thrown in – these are recipes that are special either because they introduce a particular method of preparation or baking, or because they're so completely mind-blowing I had to share them with you (like the salted-caramel-stuffed brownies on page 136 that will leave you speechless), or both.

MAKING GLUTEN-FREE BROWNIES IS EASY

Brownies on average contain the smallest amount of flour of any gluten-free bake in this book – at only 9% (see Table 3 on page 34). Compare that to bread or cookies, with percentages of around 40%, and you'll see why the nature of the flour you use is one of the less important considerations when it comes to making brownies.

For brownies, we're not interested in trying to mimic the effects of gluten – there's no room for elasticity or 'gluten development' in the world of perfectly fudgy brownies. The only possible pitfall of gluten-free brownies is that they, as any gluten-free bake, can be somewhat dry and crumbly if you use too much gluten-free flour.

That said, if you follow (and understand the reasoning behind) the recipes on the following pages, and then build on them in your brownie-baking adventures, fudgy, chocolatey perfection awaits you.

FUDGY, GOOEY, CHEWY AND NOT CAKEY, IF YOU PLEASE

The idea of what the perfect brownie should be seems to vary from person to person. While you might be a corner or edge fan, I'm a centrepiece devotee. Then, there's the question of the right balance between fudgy and gooey – my idea of a perfect brownie is on the fudgier side, with only traces of gooiness present.

Thankfully, all these things are easy to tweak, simply by adjusting the baking time – with a shorter baking time for gooey and a longer one for fudgy. The toothpick test (see page 30) is, as expected, slightly different for brownies. Don't look for a clean toothpick as you would for a cake – that would mean your brownies are overbaked. Instead, insert a toothpick into the middle of the brownies for 2 seconds and hope to find:

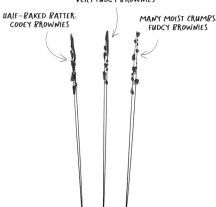

THICK HALF-BAKED
BATTER & MANY MOIST
CRUMBS:
VERY FUDGY BROWNIES

HALF-BAKED BATTER:
GOOEY BROWNIES

MANY MOIST CRUMBS:
FUDGY BROWNIES

+ **Gooey brownies** – a toothpick covered in half-baked batter. The texture of the batter should be soft and sticky to the touch, but not as runny as raw batter.
+ **Very fudgy brownies** – a toothpick with a combination of thick, half-baked batter and many moist crumbs.
+ **Fudgy brownies** – a toothpick covered in many moist crumbs.

I recommend you start testing the brownies a few minutes before the recommended baking time, and take note of how the toothpick looks after each test. Eventually, you'll know by heart what kind of toothpick corresponds to what kind of brownie.

On the other hand, the chewiness of a brownie is mostly related to the amount of brown sugar in the recipe and how long you keep the brownies in a closed container after baking. The higher the amount of brown sugar (relative to white caster sugar) and the longer you keep your brownies in a closed container, the chewier their texture.

Of course, there's that other brownie texture – cakey. I don't know about you, but if I want a cakey, chocolatey treat, I'll just have some chocolate cake. I would even go as far as to say that a cakey brownie is less a type of brownie and more an incorrectly baked brownie – so you definitely won't find a cakey brownie recipe in this book. (If you happen to be a cakey brownie fan, I'm sorry.)

In fact, avoiding a cakey brownie texture is the reason why you won't see any raising agents in my brownie recipes. Baking powder and bicarbonate of soda give brownies an open-textured crumb, which rather destroys the effect of adjusting all the other ingredients and methods to give fudgy or gooey brownies.

THE STAR OF THE SHOW: CHOCOLATE

You know how it's always said that, for cooking, you should use a wine you actually like to drink? Unsurprisingly, the same is true for chocolate. When you are making brownies, always use a high-quality chocolate, the flavour of which you really, really enjoy.

In my brownie recipes, I recommend the chocolate cocoa percentages that you should use for the most delicious results – usually in the 60–70% cocoa-solids range. You can definitely go higher, if you want a deeper chocolate flavour, but be careful when you go lower. I wouldn't recommend using a milk chocolate with cocoa solids below 50%.

Because many brownie recipes rely on whisking together sugar and eggs for their texture and appearance, you need a certain (usually quite high) amount of sugar (sugar also helps to keep the brownies nice and fudgy). If you use both the amount of sugar I've suggested and a sweeter chocolate with a lower cocoa-solids percentage, you'll end up with brownies that are almost sickly sweet. That said, if you have only milk or semi-sweet chocolate to hand, you can temper the sweetness with a slightly larger amount of cocoa powder.

But: what if you don't have any chocolate to hand and yet you really, really want a brownie? Worry not, I've got it covered – the fudgy cocoa brownie recipe on page 130 requires no actual chocolate, while being perfectly fudgy, terribly indulgent and oh-so satisfying.

THE ELUSIVE SHINY TOP

The mystery that is the paper-thin, shiny brownie top eluded me for a very long time. Browsing through the recipes online and in numerous cook books, there seemed to be no consensus about how to achieve it. Some bakers swore by whisking together eggs and sugar, others by the dissolve-sugar-in-butter method (more on that outlandish claim below). Some claimed that cocoa-only recipes are the secret, others that the addition of chocolate chips into the brownie batter makes absolutely all the difference.

There was, however, one thing many of these recipes had in common. Their results were surprisingly (and disappointingly) inconsistent. Make them one day, the brownie top was gloriously shiny. The very next day, with a new batch of brownies, following the same recipe… the paper-thin, glossy crust was nowhere in sight.

That sort of inconsistency and lack of reproducibility was rather annoying – when I repeat an experiment (even if it's a chocolatey, brownie experiment) under seemingly unchanged conditions, I expect the same outcome every single time.

My problems were two-fold: the quite popular (and scientifically somewhat bogus) dissolve-sugar-in-butter method just didn't work for me, instead producing a dull, hardly there crust. Whisking together the eggs and sugar, on the other hand, resulted in crackly top brownies (see page 129) – which, while amazing and delicious, weren't what I was going for. It was clear that how and where the sugar dissolves play central roles in this shiny-top conundrum. But, that was pretty much the only thing that was clear.

The breakthrough came as I was analysing the crust of the crackly top brownies. (Yes, I like to get up close and personal to my bakes – and, yes, there's often a magnifying glass involved. I have no regrets.) You see, I noticed that it actually comprises two parts: a crisp, meringue-like crust and, on top of it, an extremely thin, shiny layer. That shiny layer was the elusive shiny crust I was working towards all this time. So, it was there already. Now for the challenge of isolating it and creating it without the accompanying crisp, meringue-like crust.

In the end this final step wasn't much of a challenge at all. As it turns out the meringue-like crust forms as the brownie top dries out. Therefore, decreasing the amount of flour and increasing the amount of chocolate in the recipe (for that extra-fudgy texture), along with whisking the eggs and sugar together, produces brownies with a paper-thin, shiny top every single time.

I took this a step further, replacing caster sugar with light brown soft sugar, which brings even more moisture to the brownies and acts as an additional guarantee that the thin shiny crust will make an appearance. After these adjustments, the recipe for shiny top brownies (see page 126) is consistent, reliable and reproducible – all words that warm my science-loving, chocolate-obsessed heart.

DEBUNKING THE MYTH THAT SUGAR DISSOLVES IN BUTTER

I'll keep this short and sweet. Sugar does not dissolve in butter – that's just not how chemistry works. Sugar requires so-called 'polar solvents' (like water) to dissolve, and while butter does contain a small amount of water, it's simply not sufficient to dissolve the rather large amount of sugar you're adding into the brownie batter. Even at high temperatures, the amount of sugar dissolved in butter is downright pitiful, and definitely not enough to give you a nice, shiny brownie top. It's true that shiny-top formation is all about the controlled dissolving of sugar, but that dissolution happens in eggs. Not. In. Butter. Case closed.

OVEN TEMPERATURE: 160°C VS 180°C

Unless you're intending to make crackly top brownies, drying out the outside surface of brownies is just about the last thing you want to do. A lower oven temperature of 160°C ensures that the inside of the brownies has time to bake to the point of fudginess before the outside dries out too much.

At the same time, the lower temperature requires a slightly longer baking time, which in turn allows the shiny crust to form fully across the whole brownie – and not just at the very centre. Baking at a higher temperature for a shorter amount of time can often result in a patchy shiny top that is almost completely absent at the edges.

SHINY TOP BROWNIES

SERVES 9–12
PREP TIME 30 mins
BAKE TIME 24 mins

240g dark chocolate (60–70%
   cocoa solids), chopped
120g unsalted butter
3 eggs, at room temperature ①
210g light brown soft sugar
75g gluten-free flour blend
30g Dutch processed cocoa
   powder ②
½ teaspoon xanthan gum
½ teaspoon salt

NOTES

① *To get the shiny crust, the sugar*
must dissolve in the eggs as
much as possible, which is easier
for room-temperature eggs than
cold ones. If you forgot to bring
your eggs to room temperature,
place them in a bowl of warm
water for 5–10 minutes before
using. For the same reason,
don't add any water or milk to
the batter (like you might for a
chocolate cake). If you add water,
the sugar will preferentially
dissolve in it, and you can bid
the shiny crust a sad goodbye.

② *Dutch processed cocoa powder*
has a richer, deeper chocolate
flavour than natural cocoa
powder, which can be slightly
acidic. The relatively high
amount of cocoa powder reduces
the amount of flour you need,
making the brownies extra fudgy.

I present to you a 100% reliable way of getting the paper-thin, shiny crust on top of fudgy (or gooey, depending on your preference), intensely chocolatey brownies – every single time. There is no guesswork here, nor is there any need for meticulously measuring the temperature of your ingredients or anything complicated like that. The effectiveness of this recipe comes down to three things: keeping the amount of dry ingredients low and the amount of chocolate high, and whisking the eggs and sugar until thick and fluffy (which gives the sugar plenty of time to dissolve in the eggs). I've made these brownies hundreds of times now – and they still take my breath away.

+ Adjust the oven shelf to the middle position, preheat the oven to 160°C and line a 20cm square baking tin with baking paper cut large enough to overhang the sides of the tin (to help you remove the baked brownies).
+ In a heatproof bowl over a pan of simmering water, melt the dark chocolate and butter together. Set aside to cool until warm.
+ Using a stand mixer with the whisk attachment or a hand mixer fitted with the double beaters, whisk the eggs and sugar on medium–high speed for 5–7 minutes until pale, fluffy and about tripled in volume. The mixture should briefly pile on top of itself as it falls off the whisk.
+ Fold in the melted chocolate until only just incorporated.
+ Sift the gluten-free flour blend, cocoa powder, xanthan gum and salt into the chocolate mixture, and use a spatula to gently fold in until no clumps remain. Scrape down the sides and bottom of the bowl from time to time to prevent any unmixed patches.
+ Transfer to the lined baking tin, smooth out the top, and bake for 24–28 minutes (24 minutes for gooey brownies; 28 for fudgy) or until an inserted toothpick comes out covered in half-baked batter (gooey) or with many moist crumbs attached (fudgy).
+ Cool the brownies completely to room temperature before taking hold of the baking paper and removing them from the tin. Use a sharp knife to slice into portions (cooling first allows the crumb to set to give you clean slices with neat edges).
+ If you want to serve the brownies warm (for example, with a scoop of ice cream), reheat them in the microwave for about 15 seconds.

STORAGE
3–4 days in a closed container at room temperature.

PUT A TWIST ON IT
You can add everything from dark, milk or white chocolate chips to chopped nuts to these brownies – chopped, toasted hazelnuts work particularly well.

CRACKLY TOP BROWNIES

SERVES 9–12
PREP TIME 30 mins
BAKE TIME 30 mins

170g dark chocolate (60–70%
cocoa solids), chopped
210g unsalted butter
3 eggs, at room temperature
300g caster sugar
145g gluten-free flour blend
35g Dutch processed cocoa
powder
½ teaspoon xanthan gum
½ teaspoon salt

Take pretty much everything you know about how to get a shiny top
on brownies, turn it on its head, and you'll get these crackly top
brownies, with a crunchy, meringue-like crust and a superb fudgy
texture. Success relies on drying out the surface of the brownies by
adjusting the chocolate:butter ratio so that it is lower in chocolate,
using caster instead of light brown soft sugar, increasing the amounts
of the dry ingredients, upping the oven temperature to 180°C and
slightly increasing the baking time. You might think these steps will
yield disappointingly dry results, but the larger amount of butter
keeps them nice and fudgy. The result is an explosion of contrasting
textures, and a delicious addition to the vast world of brownies.

+   Adjust the oven shelf to the middle position, preheat the oven to
    180°C and line a 20cm square baking tin with baking paper cut
    large enough to overhang the sides of the tin (to help you remove
    the baked brownies).
+   In a heatproof bowl over a pan of simmering water, melt the dark
    chocolate and butter together. Set aside to cool until warm.
+   Using a stand mixer with the whisk attachment or a hand mixer fitted
    with the double beaters, whisk the eggs and sugar on medium–high
    speed for 5–7 minutes until pale, fluffy and about tripled in volume.
    The mixture should briefly pile on top of itself as it falls off the whisk.
+   Pour in the melted chocolate while whisking on low speed, then
    increase the speed to medium–high and continue whisking until all
    the chocolate is fully incorporated.
+   Sift the gluten-free flour blend, cocoa powder, xanthan gum and salt
    into the egg–chocolate mixture, and use a spatula to gently fold in until
    no clumps remain.
+   Transfer to the lined baking tin, smooth out the top, and bake for about
    30–35 minutes (30 minutes for gooey; 35 for fudgy) or until an inserted
    toothpick comes out with some half-baked batter attached (gooey) or
    with some moist crumbs attached (fudgy).
+   Cool the brownies completely to room temperature before taking hold
    of the baking paper and removing them from the tin. Use a sharp knife
    to slice into portions (cooling allows the crumb to set to give you clean
    slices with neat edges).
+   If you want to serve the brownies warm (for example, with a scoop
    of ice cream), reheat them in the microwave for about 15 seconds.

STORAGE
3–4 days in a closed container at room temperature.

FUDGY COCOA BROWNIES

SERVES 9–12
PREP TIME 30 mins
BAKE TIME 35 mins

375g caster sugar ①
3 eggs, at room temperature
210g unsalted butter, melted and
  cooled until warm
110g Dutch processed cocoa
  powder
90g gluten-free flour blend
¼ teaspoon xanthan gum
½ teaspoon salt

NOTES

① *You can reduce the amount of
sugar to about 300g, but in that
case the brownie top is unlikely
to have the smooth, glossy crust.
I don't recommend going below
300g, which will result in a
progressively less fudgy and
more cakey brownie.*

② *Aerating the egg and sugar
mixture (whisking until pale and
fluffy), as you would in brownies
made with actual chocolate,
gives cocoa brownies that are
cakey in texture. In actual
chocolate brownies, the chocolate
acts as a failsafe for fudginess,
so aeration is less of a concern.*

③ *Gently heating the eggs and
sugar together helps to fully
dissolve the sugar (without
cooking the eggs), which in
turn guarantees a beautiful,
paper-thin, shiny brownie crust.*

Although I'll be the first to admit that actual chocolate is pretty
essential for a good brownie, on the off-chance that you get a brownie
craving with no chocolate to hand, I don't want to leave you hanging.
This recipe gives brownies with the deep chocolate flavour, the fudgy
texture and the gorgeous glossy top – using only cocoa powder. It's
pretty high in sugar, but the bitterness of the Dutch processed cocoa
powder balances this sweetness nicely. Plus, the sugar is essential
for the formation of the shiny crust.

+ Adjust the oven shelf to the middle position, preheat the oven to
  160°C and line a 20cm square baking tin with a piece of baking
  paper cut large enough to overhang the sides of the tin (to help you
  remove the baked brownies).
+ In a heatproof bowl, mix together the sugar and eggs until only just
  combined. You can use a wooden spoon, a spatula or a whisk – but
  in the latter case, don't aerate the eggs. You don't want the mixture
  to become pale and fluffy, only to combine the eggs and sugar until
  homogeneous. ②
+ Place the egg–sugar mixture over a pan of simmering water and heat,
  stirring continuously, until the mixture reaches 28–30°C and the sugar
  is fully dissolved (you shouldn't be able to feel any graininess when you
  rub a little bit of the mixture between your fingertips). ③
+ Remove from the heat, add the butter and mix until fully incorporated.
  Sift in the cocoa powder, gluten-free flour blend, xanthan gum and salt
  and mix well to a smooth batter with no flour clumps.
+ Transfer to the lined baking tin, smooth out the top, and bake for about
  35–40 minutes (35 minutes for gooey brownies; 40 for fudgy) or until an
  inserted toothpick comes out covered in half-baked batter (gooey) or
  with many moist crumbs attached (fudgy).
+ Cool the brownies completely to room temperature before taking hold
  of the baking paper and removing them from the tin. Use a sharp knife
  to slice into portions (cooling first allows the crumb to set to give you
  clean slices with neat edges).
+ If you want to serve the brownies warm (for example, with a scoop
  of ice cream), reheat them in the microwave for about 15 seconds.

STORAGE
3–4 days in a closed container at room temperature. The brownies will
get progressively fudgier and chewier the longer you keep them like this.

GRAIN-FREE + DAIRY-FREE BROWNIES

SERVES 9–12
PREP TIME 30 mins
BAKE TIME 18 mins

Don't let these brownies fool you – they might be gluten-, dairy- and (mostly) refined-sugar-free, and therefore could be considered healthy-ish, but they couldn't be more indulgent if they tried. With a luscious, gooey texture, an intense chocolate flavour and a thin, crackly top, they tick every single box.

125g dark chocolate (60–70% cocoa solids), chopped ①
50g coconut oil
3 eggs, at room temperature
100g coconut sugar
90g almond flour ②
30g Dutch processed cocoa powder
¼ teaspoon salt
175g dark chocolate chips (optional)

NOTES

① *If you prefer to use a dark chocolate with 80%+ cocoa solids (which contains less refined sugar), increase the coconut sugar to 175g.*

② *If you wish to omit the almond flour (if you are allergic to almonds, for example), you can substitute it with 75g gluten-free flour blend, or 90g of another type of finely ground nuts, such as hazelnuts.*

③ *Whisking the eggs and sugar together until pale and voluminous is the secret to the delicious, crackly brownie crust.*

+ Adjust the oven shelf to the middle position, preheat the oven to 180°C and line a 20cm square baking tin with a piece of baking paper cut large enough to overhang the sides of the tin (to help you remove the baked brownies).
+ In a heatproof bowl over a pan of simmering water, melt the dark chocolate and coconut oil together. Set aside to cool until warm.
+ Using a stand mixer with the whisk attachment or a hand mixer fitted with the double beaters, whisk the eggs and coconut sugar on medium–high speed for 5–7 minutes until pale, fluffy and about tripled in volume. ③
+ Slowly drizzle in the melted chocolate while whisking on low speed until all the chocolate is fully incorporated.
+ Sift the almond flour, cocoa powder and salt into the egg–chocolate mixture, and use a spatula to gently fold in until no flour clumps remain.
+ Stir most of the chocolate chips (if using) into the brownie batter, then transfer the batter into the lined baking tin. Smooth out the top and sprinkle over the reserved chocolate chips.
+ Bake for 18–20 minutes (18 minutes for gooey brownies; 20 for fudgy) or until an inserted toothpick comes out covered in half-baked batter (gooey) or with moist crumbs attached (fudgy).
+ Cool the brownies completely to room temperature before taking hold of the baking paper and removing them from the tin. Use a sharp knife to slice into portions (cooling first allows the crumb to set to give you clean slices with neat edges).
+ If you want to serve the brownies warm (for example, with a scoop of ice cream), reheat them in the microwave for about 15 seconds.

STORAGE
3–4 days in a closed container at room temperature.

WHITE CHOCOLATE BROWNIES

SERVES 9–12
PREP TIME 30 mins
BAKE TIME 45 mins

210g white chocolate, chopped
60g unsalted butter
3 eggs, at room temperature
150g caster sugar
120g gluten-free flour blend
½ teaspoon xanthan gum
½ teaspoon salt

Contrary to the belief that white chocolate brownies and blondies are the same, in fact blondies don't contain any chocolate (unless you add chocolate chips), and are far closer to cookie bars than to brownies. The secrets to the perfect white chocolate brownie (fudgy, not overly sweet, of a pretty off-white colour with a crisp, glossy, golden crust) are threefold: slightly reducing the amount of sugar compared to dark chocolate brownies, whisking the eggs and sugar until pale and fluffy, and separately melting the butter and the white chocolate. The latter is especially important, as white chocolate is likely to split when melted with butter. In dark chocolate, cocoa powder stabilises cocoa butter in the presence of small amounts of water, but as there are no cocoa solids in white chocolate, stabilising can't happen. Butter contains anywhere from 15 to 17% water, which is enough to cause problems.

+ Adjust the oven shelf to the middle position, preheat the oven to 160°C and line a 20cm square baking tin with a piece of baking paper cut large enough to overhang the sides of the tin (to help you to remove the baked brownies).
+ Melt the white chocolate in a heatproof bowl over a pan of simmering water, then set aside to cool until needed. Separately, melt the butter and cool until warm.
+ Using a stand mixer with the whisk attachment or a hand mixer fitted with the double beaters, whisk the eggs and sugar on medium–high speed for 5–7 minutes until pale, fluffy and about tripled in volume. The mixture should briefly pile on top of itself as it falls off the whisk.
+ Fold in the melted chocolate, followed by the butter.
+ Sift in the gluten-free flour blend, xanthan gum and salt, and use a spatula to gently fold in until no flour clumps remain.
+ Transfer to the lined baking tin, smooth out the top, and bake for about 45–50 minutes or until an inserted toothpick comes out with half-baked batter or some moist crumbs attached. If the top browns too quickly, cover with a sheet of foil (shiny side up) and bake until done.
+ Cool the brownies to room temperature, before taking hold of the baking paper and removing them from the tin. Use a sharp knife to slice into portions (cooling first allows the crumb to set to give you neat slices).
+ If you want to serve the brownies warm (for example, with a scoop of ice cream), reheat them in the microwave for about 15 seconds.

STORAGE
3–4 days in a closed container at room temperature.

PUT A TWIST ON IT
Raspberry white chocolate brownies – Add some fresh raspberries, the tartness of which will balance out the sweetness of the white chocolate.

SALTED-CARAMEL-STUFFED BROWNIES

SERVES 9–12
PREP TIME 1 hour
COOK TIME 15 mins
BAKE TIME 30 mins

BROWNIES

240g dark chocolate (about
   70% cocoa solids), chopped
120g unsalted butter
3 eggs, at room temperature
210g light brown soft sugar
75g gluten-free flour blend
30g Dutch processed cocoa
   powder
½ teaspoon xanthan gum
½ teaspoon salt

CARAMEL LAYER

180g double cream
30g unsalted butter
80g golden syrup
100g caster sugar
¼–½ teaspoon salt, to taste

If there is such a thing as a food crush, this is mine. These caramel brownies are simply amazing on every single level. First, there's the combination of textures: fudgy, slightly chewy brownies with gooey caramel oozing from the centre. Then, there's the melding of flavours: the slight bitterness of the chocolate with the sweet–salty caramel deliciousness. Every bite feels like a sensory explosion, and the recipe honestly isn't much more complicated than a standard brownie recipe. But the result? Life changing. (And, trust me, I don't make such a claim lightly.)

BROWNIES

+ Line a 20cm square baking tin with a piece of baking paper cut large enough to overhang the sides of the tin (to help you remove the baked brownies).
+ In a heatproof bowl over a pan of simmering water, melt the dark chocolate and butter together. Set aside to cool until warm.
+ Using a stand mixer with the whisk attachment or a hand mixer fitted with the double beaters, whisk the eggs and sugar on medium–high speed for 5–7 minutes until pale, fluffy and about tripled in volume. The mixture should briefly pile on top of itself as it falls off the whisk.
+ Fold in the melted chocolate.
+ Sift in the gluten-free flour blend, xanthan gum, cocoa powder and salt, and use a spatula to gently fold in until no flour clumps remain.
+ Pour half the brownie batter (about 400g) into the lined baking tin, smooth it out and refrigerate until needed. ① Set aside the other half at room temperature.

CARAMEL LAYER + ASSEMBLING THE BROWNIES

+ Adjust the oven shelf to the middle position and preheat the oven to 160°C.
+ In a saucepan, bring the double cream and 15g of the butter to the boil. Remove from the heat and set aside until needed, making sure the mixture stays warm.
+ In a clean saucepan, combine the golden syrup and sugar, and cook over a medium–high heat, stirring frequently, until it reaches 145°C. Remove from the heat, add the cream mixture and mix until combined, then heat again, stirring continuously, to 112–116°C. ② Immediately remove from the heat and stir in the salt and the remaining butter.
+ Stirring occasionally, cool the caramel to 80°C – at this temperature, it's unlikely to melt the bottom brownie layer while still being pourable and spreadable. Pour this over the chilled brownie layer in the baking tin and spread it out, leaving about a 1cm border around the edges (to prevent the caramel from leaking out too much during baking).
+ Spoon the other half of the brownie batter over the caramel and gently spread it out to completely cover the caramel layer.
+ Bake for about 30–35 minutes until an inserted toothpick comes out covered in half-baked batter or with a few moist crumbs attached.
+ Cool the brownies completely to room temperature before taking hold of the baking paper and removing them from the tin. Use a sharp knife to slice into portions. (Cooling first allows the crumb to set to give you neat slices with clean edges. Although these brownies, served warm from the oven, will have caramel invitingly oozing out of them, without cooling they are a nightmare to cut.)
+ If you want to serve the brownies warm (for example, with a scoop of ice cream), reheat them in the microwave for about 5–10 seconds to make the caramel soft again.

STORAGE
3–4 days in a closed container at room temperature.

NOTES

① *Chilling the bottom brownie layer makes it sturdier and less likely to melt or displace as you pour the caramel on top. The other half of the brownie batter should remain runny enough at room temperature so that you can easily spread it on top of the caramel layer.*

② *Between 112 and 116°C, the caramel is in the so-called 'softball' stage. This is the optimal temperature range to give a caramel layer that, despite being heat-treated again during baking, is luxuriously soft even once the finished brownies have cooled completely. To test for the softball stage, drop a small amount of the caramel into a glass of cold water and then scoop it out – it should hold its shape but be very soft when squeezed. (You don't need to do this if you're using a sugar or digital thermometer.)*

PHOTOGRAPHY OVERLEAF ⟶

CHEESECAKE BROWNIES

SERVES 9–12
PREP TIME 45 mins
BAKE TIME 50 mins
CHILL TIME 30 mins

Let's get one thing out of the way: brownies and cheesecake are a match made in heaven. Baking the brownies separately until 90% done before adding the cheesecake guarantees two distinct layers: a creamy cheesecake with a fudgy brownie underneath. Mixing the cheesecake batter on the lowest speed of the stand mixer ensures it remains thick, so is less likely to penetrate and displace the brownie mixture as a runny, thin batter might.

BROWNIE LAYER

160g dark chocolate (60–70% cocoa solids), chopped
80g unsalted butter
140g light brown soft sugar
2 eggs, at room temperature
50g gluten-free flour blend
20g Dutch processed cocoa powder, plus extra to serve
¼ teaspoon xanthan gum
½ teaspoon salt

CHEESECAKE LAYER

300g full-fat cream cheese, at room temperature
60g full-fat plain yoghurt or soured cream, at room temperature
75g caster sugar
1 tablespoon cornflour
1 egg, at room temperature
½ tablespoon lemon juice

NOTE

① *Whisking the eggs and sugar together would create a thin, shiny crust on top of the brownies – and you don't want that in this case: a shiny crust may peel off and mix into the cheesecake batter as you pour it on top of the brownies. And, besides, you're going to cover the top of the brownie with cheesecake anyway.*

BROWNIE LAYER

+ Adjust the oven shelf to the middle position, preheat the oven to 160°C and line a 20cm square baking tin with a piece of baking paper cut large enough to overhang the sides of the tin (to help you remove the baked brownies).
+ In a heatproof bowl over a pan of simmering water, melt the chocolate and butter together. Remove from the heat and allow to cool slightly. Add the sugar and mix well. Add the eggs and mix until combined. ①
+ Sift in the gluten-free flour blend, cocoa powder, xanthan gum and salt, and mix well until no flour clumps remain.
+ Transfer to the baking tin, smooth out the top, and bake for 20 minutes. While the brownies are baking, prepare the cheesecake layer.

CHEESECAKE LAYER

+ To make the cheesecake mixture, you can use a stand mixer on the lowest speed setting and fitted with the paddle attachment, or use a balloon whisk (mixing, though, not whisking) to prepare it by hand (avoid a hand-held electric mixer, which will incorporate too much air).
+ Mix the cream cheese and yoghurt together for about 1 minute until smooth. Add the sugar and cornflour and mix well. Add the egg and lemon juice and combine to a smooth batter.

ASSEMBLING THE BROWNIES

+ Once the brownie mixture has been baking for 20 minutes, remove it from the oven. From a height of about 2–3cm above the tin (so as not to displace the part-baked brownie batter), gently pour the cheesecake on top of the brownie, taking care not to combine the two layers.
+ Smooth out the top, and return to the oven for about 30–32 minutes until the edges are fully set and the middle is still very slightly wobbly.
+ Cool for about 1–2 hours at room temperature (or at least 30 minutes in the fridge), before taking hold of the baking paper and removing the brownie from the tin. Use a sharp knife to slice into portions (cooling first allows the crumb to set to give you neat slices – an hour in the fridge gives extra-clean edges). Dust with cocoa powder to serve.

STORAGE

3–4 days in a closed container in a cool, dry place.

TOASTED MARSHMALLOW BROWNIES

SERVES 9–12
PREP TIME 45 mins
BAKE TIME 24 mins

Fudgy, dark chocolate brownies and fluffy, marshmallowy Swiss meringue, toasted to caramelised perfection: if it sounds amazing, imagine how actually amazing that very first bite will be. These brownies are incredibly easy to prepare, with mouthwatering results.

BROWNIES

240g dark chocolate (60–70% cocoa solids), chopped
120g unsalted butter
210g light brown soft sugar
3 eggs, at room temperature
75g gluten-free flour blend
30g Dutch processed cocoa powder
½ teaspoon xanthan gum
½ teaspoon salt

SWISS MERINGUE

3 egg whites
150g caster sugar
¼ teaspoon cream of tartar
1 teaspoon vanilla bean paste

NOTE

① *Whisking the eggs and sugar together would create a thin, shiny crust on top of the brownies – and you don't want that in this case: a shiny crust may peel off and mix into the Swiss meringue. And, besides, you're going to cover the top of the brownie with meringue anyway.*

BROWNIES

+ Adjust the oven shelf to the middle position, preheat the oven to 160°C and line a 20cm square baking tin with a piece of baking paper cut large enough to overhang the sides of the tin (to help you remove the baked brownies).
+ In a heatproof bowl over a pan of simmering water, melt the dark chocolate and butter together. Set aside to cool until warm.
+ Add the sugar to the melted chocolate and mix well. Add the eggs and mix until just combined. ①
+ Sift in the gluten-free flour blend, cocoa powder, xanthan gum and salt, and mix well until no flour clumps remain.
+ Transfer to the lined baking tin, smooth out the top, and bake for 24–28 minutes (24 minutes for gooey brownies; 28 for fudgy) or until an inserted toothpick comes out covered in half-baked batter (gooey) or with many moist crumbs attached (fudgy).
+ Cool the brownies to room temperature in the tin.

SWISS MERINGUE

+ Mix the egg whites, sugar and cream of tartar in a heatproof bowl over a pan of simmering water. Stir continuously until the meringue mixture reaches 65°C and the sugar has dissolved.
+ Remove the meringue from the heat, transfer it to a stand mixer with the whisk attachment or use a hand mixer fitted with the double beaters, and whisk on medium–high speed for 5–7 minutes until it forms stiff, glossy peaks. Add the vanilla paste and whisk again to combine.

ASSEMBLING THE BROWNIES

+ Spoon the meringue on top of the cooled brownies and, using a spoon or a spatula, create swirls and peaks.
+ Use a kitchen blow torch to toast the meringue. (Alternatively, place it under a hot grill for about 20 seconds until nicely caramelised.)
+ Allow to cool before taking hold of the baking paper and removing from the tin. Use a sharp knife to slice into portions (cooling first allows the crumb to set to give you neat slices). If the meringue sticks to your knife, lightly grease it with a neutral oil before cutting.

STORAGE

3–4 days in a closed container at room temperature.

TRIPLE CHOCOLATE BROWNIES

SERVES 9–12
PREP TIME 45 mins
BAKE TIME 24 mins
CHILL TIME 35 mins

When it comes to brownies and chocolate, sometimes more is just more: more delicious, more mesmerising, more holy-cow-this-is-absolutely-epic. And these triple chocolate brownies are nothing if not 'more', with white chocolate chips dotted around intensely fudgy, chewy brownies and a luscious layer of glossy chocolate ganache swirled on top.

BROWNIES

240g dark chocolate (60–70% cocoa solids), chopped
120g unsalted butter
210g light brown soft sugar
3 eggs, at room temperature
75g gluten-free flour blend
30g Dutch processed cocoa powder
½ teaspoon xanthan gum
½ teaspoon salt
100–125g white chocolate chips

CHOCOLATE GANACHE

200g dark chocolate (60–70% cocoa solids), chopped
275g double cream

NOTE

① *Whisking the eggs and sugar together would create a thin, shiny crust on top of the brownies – and you don't want that in this case: a shiny crust may peel off and mix into the ganache. And, besides, you're going to cover the top of the brownie with ganache anyway.*

BROWNIES

+ Adjust the oven shelf to the middle position, preheat the oven to 160°C and line a 20cm square baking tin with a piece of baking paper cut large enough to overhang the sides of the tin (to help you remove the baked brownies).
+ In a heatproof bowl over a pan of simmering water, melt the dark chocolate and butter together. Set aside to cool until warm.
+ Add the sugar to the melted chocolate and mix well. Add the eggs and mix until just combined. ①
+ Sift in the gluten-free flour blend, cocoa powder, xanthan gum and salt, and mix well until no flour clumps remain. Fold in the white chocolate chips until evenly distributed.
+ Transfer to the lined baking tin, smooth out the top, and bake for 24–28 minutes (24 minutes for gooey brownies; 28 for fudgy) or until an inserted toothpick comes out covered in half-baked batter (gooey) or with many moist crumbs attached (fudgy).
+ Cool the brownies in the baking tin to room temperature.

CHOCOLATE GANACHE

+ Place the dark chocolate in a heatproof bowl.
+ In a saucepan, bring the double cream just to the boil, then pour it over the chocolate. Allow to stand for 4–5 minutes, then stir together into a smooth, glossy ganache.
+ Refrigerate the ganache for about 20 minutes, stirring frequently, until spreadable. Spread the ganache on top of the cooled brownies, creating swirls with an offset spatula or the back of a spoon.
+ Allow the ganache to set at room temperature for about 30 minutes or in the fridge for about 15 minutes, before taking hold of the baking paper and removing the brownie from the tin. Use a sharp knife to slice into portions (cooling first allows the brownie crumb and the ganache to set to give neat slices).

STORAGE

3–4 days in a closed container at room temperature.

COOKIES
+
BARS

If you take a peek into my fridge or freezer, likelihood is you'll find a batch of (usually chocolate chip) cookie dough, ready to be popped into the oven at a moment's notice. Part of the reason is that I seem to be experimenting with new cookie recipes pretty much all the time. But more importantly, who doesn't love a warm cookie, still a bit gooey in the middle, 15 minutes after you think to yourself – I could really do with a little indulgent treat round about now?

The magic of cookies lies in the fact that, with just one basic cookie-dough recipe, you can go on and create thousands of completely different cookies, just by playing around with the flavours, baking times, oven temperatures and flour quantities, and baking them as regular-sized cookies or in a cast-iron skillet, or stuffing them with everything from caramels to marshmallows or a dollop of hazelnut spread.

And don't even get me started on the textures! While brownies can be fudgy or chewy or gooey, cakes moist or fluffy or a bit denser and rich, cookies can be all of that and more. Chewy, fudgy, gooey, crispy, crumbly, melt-in-the-mouth, cakey or even flaky – the world of cookies is one where texture isn't an afterthought or a side story, but rather plays a central role.

THE TYPES OF COOKIE + BAR IN THIS CHAPTER

This chapter focuses almost exclusively on a variety of fundamental gluten-free cookies, biscuits and bars. Through these, I want you to gain the knowledge you need to experiment beyond these pages. For example, instead of giving you five (or ten, or twenty) versions of a chocolate chip cookie, there's just one basic recipe. It explains the science behind the ingredients and methods, so that you can make your own adjustments to tweak the texture to your own ideal (gooey, chewy or crispy – or a delicious combination of all three). That said, to kick-start your experimentation, many recipes also have ideas for 'twists' that will help you reach new heights of cookie awesomeness.

THE ROLE OF GLUTEN

Cookies, along with bread, for the most part contain the largest amount of flour – around 40% of all ingredients (see Table 3 on page 34). The effect of this, of course, is that gluten plays a more important role in the world of cookies than many other types of bake.

On the one hand, gluten prevents the cookies from being both too dry and too crumbly. In gluten-free baking, then, you have to adjust the ingredient quantities and add xanthan gum, otherwise you can end up with cookies that crack on rolling, can't be shaped into balls and are too crumbly to pick up – basically, with cookies that are rather impossible to make, shape or eat. (Don't worry, the recipes in here are nothing like that.)

On the other hand, however, gluten can cause problems for wheat-based cookie baking. Overworking the dough can result in gummy cookies that have none of that wonderful melt-in-the-mouth texture, but feel tough and unpleasant. With gluten-free baking, you don't have to worry about overmixing or overkneading the dough, as there is no gluten to develop in the first place.

So, while gluten is important in 'regular' cookie making, it's quite easy to manage its absence, and you can re-use cookie-dough trimmings as many times as you want (provided you don't mix in too much flour in the process), as there's essentially no such thing as overworking gluten-free cookie dough.

HOW THE TYPE OF SUGAR AFFECTS COOKIE TEXTURE

Before we get to the nitty-gritty of (gluten-free) cookie making, a small disclaimer. The following words and phrases will be scattered throughout the next few pages: 'in general', 'sometimes', 'no clear trend' and 'too many variables'.

The short of it is that while I can speak in general terms about the science of baking cookies and the effects of various ingredients on their texture and appearance, sometimes there's just no clear trend because we're trying to find a trend where there are simply one (or ten) too many variables. (See what I did there?)

Keeping that in mind, we can still draw some general conclusions that will guide our cookie adventures. Let's start with sugar – specifically, with the difference between white caster sugar (or, more rarely granulated sugar) and light or dark brown soft sugar.

If we look at their composition (very simplistically), brown soft sugar is just white sugar with varying amounts of molasses added. This makes it richer in flavour, slightly acidic in pH and more hygroscopic (it absorbs more moisture from the air) than white sugar.

What does this mean for our cookies? Well, in general, brown sugar makes cookies more flavourful, and moister, gooier or chewier, depending on the specific cookie and the way it's baked. Looking at chocolate chip cookies (see page 152), increasing the brown:white sugar ratio takes the cookies from crispy to gooey. In the fudgy brownie cookies (see page 157), the 1:1 ratio of brown:white sugar ensures they keep their shape, and have a fudgy centre and a beautiful thin, shiny crust.

Any raising agents in the cookies will gain a boost in their activity thanks to the slightly acidic pH of brown sugar, although other ingredients present in the dough will usually overshadow this effect.

THE NEED FOR RAISING AGENTS

The question of whether or not we should add raising agents to cookie dough has to be handled on a cookie by cookie basis. If you're looking to make cut-out cookies that need to hold their shape during baking – without spreading or losing their neat, straight edges – don't add any raising agents. (Usually.)

In fact, I like to go a step further and use a mixing method that incorporates the minimal amount of air possible, usually by hand-kneading the cookie dough or using a wooden spoon. That's because the aeration achieved by whisking butter, sugar and eggs can have the same effect as chemical leavening agents – the trapped air expands as it hits the heat of the oven, causing the cookies to spread. We can see this in the difference between the one-bowl sugar cookies (see page 158) and gingerbread cookies (see page 173). The sugar cookies, prepared by the kneading method and containing no raising agents, maintain their shape to within the millimetre and have beautifully straight, sharp edges. On the other hand, the gingerbread cookies are prepared by whisking butter, sugar and eggs together, and they contain bicarbonate of soda. Accordingly, they spread slightly – not so much that they lose their shape (we want to keep any Christmas cookie shapes recognisable), but enough to have slightly rounded edges.

That said, in the digestives (see page 162) and rosemary crackers (page 174), the baking powder doesn't drastically change the shape of

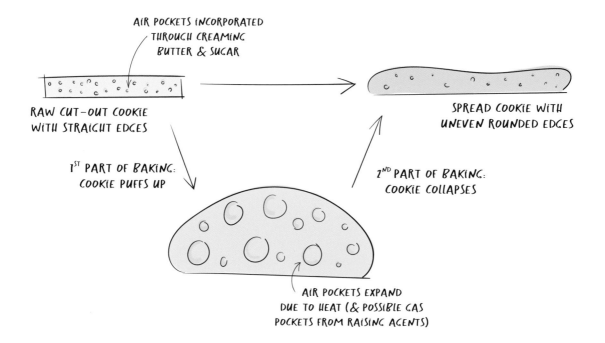

AIR POCKETS INCORPORATED THROUGH CREAMING BUTTER & SUGAR

RAW CUT-OUT COOKIE WITH STRAIGHT EDGES

SPREAD COOKIE WITH UNEVEN ROUNDED EDGES

1ST PART OF BAKING: COOKIE PUFFS UP

2ND PART OF BAKING: COOKIE COLLAPSES

AIR POCKETS EXPAND DUE TO HEAT (& POSSIBLE GAS POCKETS FROM RAISING AGENTS)

the biscuits (yes, I know these are also cut-out cookies – remember what I said about trends?), but rather opens up their texture to make them pleasantly crumbly.

If we're not dealing with cut-out cookies, the answer to the raising agents' question becomes even less clear cut. In chocolate crinkle cookies (see page 170) and fudgy brownie cookies (page 157), raising agents are crucial to their crackly appearance, and they also give chocolate chip cookies (page 152) their characteristic fudgy texture and ripply appearance, as the cookies puff up and then collapse during baking (especially if you bang the baking sheet on the counter a couple of times throughout the baking process). And then there are the peanut butter sandwich cookies (see page 169) and thumbprint cookies (page 165), where adding raising agents would lead to nothing but trouble, with the cookies spreading and cracking all over the place.

To conclude: raising agents play an important role in the world of cookies, and their presence or absence can be the very thing that makes or breaks a cookie. So, trust and follow the recipe. (And forget finding clear and comprehensive trends. Just enjoy the cookies.)

The air pockets (from creaming together the butter and sugar) and raising agents in the cookie dough cause the cookies to puff up and spread in the oven – something you want to avoid with cut-out cookies.

YOLKS VS WHITES

In general, egg yolks act to preserve the shape of the cookies (for cut-out cookies) or to make their texture chewier, fudgier or gooier (for chocolate chip cookies and similar). In contrast, egg whites often act similarly to raising agents, making the cookies rise and puff up in the oven, giving them a cakier texture.

THE UNDER-APPRECIATED INGREDIENT

Spoiler alert: it's salt. When you browse through the recipes in this chapter, you'll notice most of the cookies contain anywhere from ¼ to (a whopping) 1 teaspoon of salt. The reason is that salt is a natural flavour enhancer and it makes the flavour of cookies pop, without making the cookies taste savoury. If you've ever eaten an underwhelming chocolate chip cookie (even if its texture was spot on), you'll know what I'm talking about. We often concentrate on the sugar aspect of cookies and biscuits, and allow salt to fall aside – big, big mistake.

The science behind the role of salt in cookies is actually really interesting: salt intensifies our body's ability to taste the sweetness of sugar. That's why salted caramel tastes absolutely amazing and a sprinkling of flaky sea salt on top of chocolate chip or brownie cookies takes them to the next level. So, don't skimp on the salt – a generous pinch added to cookie dough will make a world of difference.

IS CHILLING COOKIE DOUGH REALLY THAT IMPORTANT?

Yes, but if you're impatient (like yours truly), you can often skip it with minimal sacrifice on the flavour, texture and appearance front.

Chilling pretty much any cookie dough allows the flavours to develop further and meld together into something particularly delicious. This is especially true when it comes to chocolate chip cookies, which is why it's a really good idea to make a big batch, bake only a small part of it and keep the rest in the fridge for when the cookie craving strikes.

That said, I'm not a fan of cookie recipes that chill the dough just to make it easier to handle and keep its shape during baking. There are, of course, a few exceptions to this: the aforementioned chocolate chip cookies, as well as the chocolate crinkle cookies – those start out resembling brownie batter and really need that chilling time. But, those aside, in my experience cookie dough that needs chilling heats up in no time in a warm kitchen (resulting in a mess) and the cookies will still lose much of their shape in the oven (also resulting in a mess).

That's why many of my recipes are of the no-chill variety. The one-bowl sugar cookies (see page 158), gingerbread cookies (page 173) and even shortbread biscuits (page 161) are all among those you can handle, roll, shape, cut and bake straight away without refrigeration, without any problems. The only situation where you might want to chill them is if your kitchen is extra warm and is making the butter in the dough melt at an accelerated rate. Otherwise, proceed without chilling.

This also means that the time between getting that sneaky cookie craving creep up on you and popping the first (buttery, warm, possibly chock-full of melty chocolate chips) cookie into your mouth is significantly shortened. And I really can't see a downside to that.

CHOCOLATE CHIP COOKIES

MAKES 16
PREP TIME 30 mins
CHILL TIME 1 hour 30 mins
BAKE TIME 3 x 12 mins

100g unsalted butter, melted
  and cooled
70g caster sugar
120g light brown soft sugar ①
1 egg, at room temperature
1 egg yolk, at room temperature ②
½ teaspoon vanilla bean paste
220g gluten-free flour blend
½ teaspoon xanthan gum
1 teaspoon baking powder
½ teaspoon bicarbonate of soda
½ teaspoon of salt
50g dark chocolate chips
50g dark chocolate (60–70% cocoa
  solids), roughly chopped
flaky sea salt, for sprinkling
  (optional)

Crispy. Chewy. Fudgy. Gooey. With a scattering of chocolate chips, or absolutely loaded with chocolate chunks. Maybe with a sprinkling of sea salt, or a handful of chopped toasted pecans. Everyone has an opinion about what a chocolate chip cookie absolutely should be. And even considering all these opinions and possibilities, this simple recipe offers something for everyone. With a few small tweaks and adjustments, you can get from gooey to crispy effortlessly.

+ Using a stand mixer with the paddle attachment or a hand mixer fitted with the double beaters, cream the butter and both sugars together until pale and fluffy.
+ Add the egg, egg yolk and vanilla paste, and mix until fully combined.
+ Sift together the gluten-free flour blend, xanthan gum, baking powder, bicarbonate of soda and salt. Add the dry ingredients to the wet in two batches, mixing well after each addition.
+ Stir in the chocolate chips and 25g of the chopped chocolate to evenly distribute, then refrigerate the cookie dough for about 1 hour or until firm enough to scoop.
+ Scoop the chilled cookie dough using a 2-tablespoon cookie or ice-cream scoop to give 16 balls of dough. Place the balls on a baking sheet, cover with cling film and refrigerate for at least 30 minutes.
+ Adjust the oven shelf to the middle position, preheat the oven to 180°C and line a baking sheet with baking paper.
+ In batches (at most, 6 at a time), place the chilled cookie balls on the lined baking sheet, leaving plenty of space between them.
+ Place in the oven. After 8 minutes, take the baking sheet out of the oven, tap it 5–7 times on the work surface to flatten the cookies, and divide the remaining chopped chocolate equally between the tops. ③ (Omit the tapping if you want thicker cookies.) At this point you can also 'correct' the shape of the cookies by nudging their sides with a round cutter slightly larger than the diameter of the cookies.
+ Return the cookies to the oven for a further 4–9 minutes (giving 12–17 minutes' baking in total), depending on your preferred cookie texture and consistency. ④ You can tap the baking sheet again at 2-minute intervals until the cookies are baked, to give them an extra rippled, crinkled appearance, if you wish.
+ Once the cookies are baked, remove them from the oven and, while hot, correct their shape again, using a round cutter.
+ Allow to cool on the baking sheet for 2–3 minutes, then transfer to a wire rack to cool completely. Repeat the baking instructions for the remaining balls of cookie dough, then once all the cookies are baked and cooled, sprinkle with sea salt, if you wish.

STORAGE

1 week in a closed container at room temperature. Note that the cookies will get a bit softer and chewier with time. Alternatively, you can keep raw, unbaked balls of cookie dough in an airtight container in the fridge for about 1 week, or in the freezer for 1–2 months. Bake the cookie balls directly from chilled or frozen, just adding 2–3 minutes to the baking time to compensate. Note that the longer the cookie balls are in the fridge or freezer, the less they will spread while baking (so the thicker the results).

PUT A TWIST ON IT

Tahini chocolate chip cookies – Reduce the amount of butter to 75g and add 100g tahini paste.

Pecan & white chocolate chip cookies – Roast a handful of pecans, chop them up and pair them with white chocolate.

Table 5: The effect of baking time on the texture of warm and completely cooled chocolate chip cookies.

BAKING TIME	WARM OUT OF THE OVEN	COMPLETELY COOLED
12–13 minutes	very gooey	very fudgy
14–15 minutes	fudgy	chewy
16–17 minutes	fudgy-chewy	chewy-crispy

NOTES

① *Use dark brown soft sugar (or muscovado sugar) instead of light for extra depth of flavour. In this case, increase the gluten-free flour blend to 240g, as dark sugar contains more molasses and therefore more moisture. The baked cookies will have a fudgier or gooier texture, depending on the baking time.*

② *The additional egg yolk increases the fat content of the cookies giving both greater chewiness and more gooeyness. If you want the cookies to have a cakey texture, use 1 whole egg and 1 egg white instead of 1 whole egg and 1 egg yolk.*

③ *The amazing Sara Kieffer from The Vanilla Bean Blog popularised the phenomenon that is 'pan banging'.*

④ *The exact baking time will determine the texture and consistency of the cookie, both right out of the oven and a few days later. See Table 5 for the differences you can expect (note that precise times will vary according to oven size and calibration).*

PHOTOGRAPHY OVERLEAF ⟶

FUDGY BROWNIE COOKIES

MAKES 10
PREP TIME 30 mins
BAKE TIME 2 x 10 mins

160g dark chocolate (60–70%
cocoa solids), chopped
80g unsalted butter
70g light brown soft sugar ①
70g caster sugar
2 eggs, at room temperature
80g gluten-free flour blend
20g Dutch processed cocoa
powder
½ teaspoon baking powder
½ teaspoon xanthan gum
½ teaspoon salt
flaky sea salt, for sprinkling

NOTES

① *The 1:1 ratio of light brown soft
and caster sugar ensures that
the cookies don't spread too
much during baking and have a
beautiful, thin, shiny crust, while
keeping their centres fudgy.*

② *This timing is quite important –
if you whisk the eggs and sugar
for 5–7 minutes, the melted
chocolate will have just enough
time to cool down to the perfect
temperature to be folded into the
whisked eggs and so to give you
the best baked results.*

③ *You can use a spoon to scoop
the cookie batter if you prefer, but
you probably won't get quite as
uniformly circular cookies. If you
use a smaller ice-cream scoop,
reduce the baking time a little.*

Of all the brownie cookie recipes I've tested, this is my favourite.
It ticks every box: decadent fudginess, a beautiful, shiny, crackly top
and intense chocolatiness. But unlike many other recipes, this one is
far less finicky and susceptible to failure (if you have a delay between
mixing and baking, say). The cookies also contain slightly less sugar
than is usual, allowing the gorgeous bittersweet chocolate to shine.

+ Adjust the oven shelf to the middle position, preheat the oven
  to 180°C and line two baking sheets with baking paper.
+ In a heatproof bowl over a pan of simmering water, melt together
  the chocolate and butter. Once melted, remove from the heat.
+ Using a stand mixer with the whisk attachment or a hand mixer
  fitted with the double beaters, whisk both sugars and the eggs on
  medium–high speed for 5–7 minutes until pale and tripled in volume. ②
  The mixture should briefly pile on top of itself as it falls off the whisk.
+ Gently fold the melted chocolate mixture into the egg mixture until only
  just incorporated.
+ Sift in the gluten-free flour blend, cocoa powder, baking powder,
  xanthan gum and salt, and use a spatula to gently fold in until only just
  incorporated and no flour clumps remain. The cookie batter will be
  fairly runny and very glossy, strongly resembling brownie batter.
+ Use a large, 3-tablespoon cookie or ice-cream scoop to transfer the
  cookie batter to the lined baking sheets, giving 10 cookies. ③ Evenly
  space them at least 2.5cm apart as they will spread during baking.
  As the batter is fairly runny, invert the scoop about 2–3cm above the
  baking sheet to form a circular mound. You should fit about 5 cookies
  per baking sheet (scoop all the cookies at once as the mixture tends
  to firm up when standing at room temperature).
+ One baking sheet at a time, bake the cookies for about 10 minutes
  or until slightly increased in size, shiny and crackly on top, and still
  slightly soft to the touch. Once the first batch is cooked, immediately
  place the second sheet in the oven.
+ Cool the cookies on the baking sheets for 5–10 minutes, then transfer
  to a wire rack to cool completely. Sprinkle with flaky salt to serve.

STORAGE
1 week in a closed container at room temperature.

PUT A TWIST ON IT
You can stuff the brownie cookies with things like hazelnut spread, peanut
butter or caramel. Spoon about 1½ tablespoons of cookie batter on to the
baking sheet in a rough circle. Place about 1–2 teaspoons of the filling in
the middle and top with a further 1½ tablespoons of cookie batter, making
sure the filling is covered. Bake as described in the recipe.

NO-FUSS, NO-CHILL, NO-SPREAD ONE-BOWL SUGAR COOKIES

MAKES 40
PREP TIME 30 mins
BAKE TIME 2 x 10 mins

360g gluten-free flour blend
200g caster sugar
1 teaspoon xanthan gum
¼ teaspoon salt
225g unsalted butter, softened
1 egg, at room temperature
1 teaspoon vanilla bean paste

NOTE

① *You want the cut-out cookies to keep their shape and neat, straight edges, so kneading the cookie dough is preferable to creaming the butter and sugar together with a mixer, which aerates the dough. In which case, the trapped air acts similarly to chemical leavening agents, causing the cookies to spread in the oven. (Note the absence of baking powder and bicarbonate of soda for the same reason.)*

As the recipe name suggests, baking cookies doesn't get much easier than this. You knead everything by hand (no food processor or mixer required) – and the final dough, although slightly on the softer side, is still firm enough to roll out, cut and bake directly, without any need to chill it. For all their simplicity, the sugar cookies hold their shape beautifully and have the most wonderful, buttery flavour and melt-in-the-mouth texture that gets better with time. In fact, the cookies reach their peak of deliciousness three or four days after baking.

+ Adjust the oven shelf to the middle position, preheat the oven to 180°C and line two baking sheets with baking paper.
+ In a bowl, whisk together the gluten-free flour blend, sugar, xanthan gum and salt. Add the butter, egg and vanilla paste, and knead everything together into a smooth, soft (but not sticky) dough. ①
+ Generously flour the work surface and roll out the cookie dough until about 3–4mm thick. Slide a long, offset spatula underneath the dough to loosen any patches that have stuck to the work surface.
+ Cut out the cookies using a 6cm cutter, re-rolling the trimmings as necessary, until you have 40 cookies. Place the cookies on the lined baking sheets about 5mm apart (the cookies don't spread significantly).
+ One baking sheet at a time, bake the cookies for about 10–12 minutes or until light golden brown around the edges.
+ Allow to cool on the baking sheets for about 10 minutes, then transfer to a wire rack to cool completely.

STORAGE

1–2 weeks in a closed container at room temperature.

PUT A TWIST ON IT

Chocolate sugar cookies – Add about 40–45g Dutch processed cocoa powder to the dry ingredients. These are wonderful half-dipped into chocolate, or sandwiched together with some chocolate ganache.

SHORTBREAD BISCUITS

MAKES 18
PREP TIME 30 mins
BAKE TIME 20 mins

150g unsalted butter, softened
75g caster sugar, plus extra for
   sprinkling
1 teaspoon vanilla bean paste
255g gluten-free flour blend
60g cornflour
¾ teaspoon xanthan gum
½ teaspoon salt

NOTES

① *Mixing with a wooden spoon or
spatula, as opposed to a whisk,
prevents aerating the dough, so
that the cookies hold their shape
during baking and don't puff up.*

② *You can roll the dough out
thinner, if you like – just shorten
the baking time. I've found that
thinner cookies (around 5mm
thick) hold their shape and neat
edges even better, but their
texture is crisper.*

The fewer the ingredients in a recipe, the more important their exact
ratios and the method by which you combine them. This is especially
true for these buttery shortbread biscuits. Too little flour and they'll
melt into a puddle; too much and you'll overpower the renowned,
buttery, shortbread flavour. Similarly, beating the sugar and butter
until pale and fluffy incorporates too much air, causing the biscuits
to spread unattractively during baking. But don't worry – this recipe
strikes the perfect balance, giving shortbread biscuits that are buttery
and sweet, with just enough salt to make their simple flavour pop,
and that hold their shape beautifully.

+ Adjust the oven shelf to the middle position, preheat the oven to 160°C
  and line a baking sheet with baking paper.
+ In a bowl, use a wooden spoon or spatula to mix together the butter,
  sugar and vanilla paste until smooth. ①
+ Sift in the gluten-free flour blend, cornflour, xanthan gum and salt.
  Mix with the spoon or spatula and then knead by hand until the
  dough comes together in a ball. (The dough will be very crumbly at
  the beginning, but will come together after a few minutes of kneading.)
+ Compress the dough together into a disc between two pieces of baking
  paper, and roll it out into a 15 x 23cm rectangle about 1cm thick. ②
+ Cut the dough into 2.5 x 7.5cm rectangles, sprinkle generously with
  caster sugar, prick with a fork or toothpick, and transfer to the baking
  sheet, spacing the rectangles about 1cm apart.
+ Bake for about 20–22 minutes until very light golden around the edges.
+ Allow to cool on the baking sheet for 10 minutes, then transfer to a wire
  rack to cool completely.

STORAGE
1–2 weeks in a closed container at room temperature.

PUT A TWIST ON IT
You can dip the baked and cooled plain shortbread biscuits into melted
chocolate, or add herbs (such as thyme or rosemary) or spices (such as
cardamom, ginger or cinnamon) directly into the cookie dough.

CHOCOLATE-COVERED DIGESTIVES

MAKES 32
PREP TIME 30 mins
BAKE TIME 2 x 14 mins

90g gluten-free flour blend
90g brown rice flour
90g gluten-free oats, coarsely
　ground
90g light brown soft sugar
1 teaspoon baking powder
½ teaspoon xanthan gum
½–1 teaspoon salt ①
165g unsalted butter, cubed and
　chilled
50g whole milk, cold
250g dark chocolate (60–70%
　cocoa solids), melted and
　tempered (optional) ②

NOTES

① *I prefer my digestives with a salty
kick, but you can add less salt
(although I wouldn't recommend
going below ½ teaspoon).*

② *Tempering the chocolate gives
a nice, glossy finish. To temper:
melt ¾ of the chocolate, remove
from the heat, then add the
remainder (called 'seeding',
this encourages the chocolate
to crystallise correctly). This
quickly cools the melted
chocolate. Reheat it to the
optimal temperature range
of 31–33°C, then use.*

③ *Depending on the coarseness
of the oats, you might need a
bit more flour or milk. The final
dough shouldn't feel sticky
and should hold its shape.*

Normally, digestive biscuits rely on wholemeal flour and malt extract, both of which contain gluten, for their characteristic flavour and texture. These gluten-free digestives use a mixture of gluten-free flour blend, brown rice flour and coarsely ground oats in combination with light brown soft sugar. They replicate the flavour and texture with great success. (Although the recipe is written for a food processor, you can easily make the biscuits by hand. Just work the butter into the dry ingredients with your fingers, then use a fork or wooden spoon to mix in the milk.)

+ Adjust the oven shelf to the middle position, preheat the oven to 180°C and line two baking sheets with baking paper.
+ Add the gluten-free flour blend, brown rice flour, oats, sugar, baking powder, xanthan gum and salt to the bowl of a food processor, and blend briefly to mix thoroughly.
+ Add the cold butter and process to a coarse-sand texture.
+ Add the cold milk and process until the dough starts coming together in a ball (it can be slightly on the crumblier side depending on how coarsely you've ground your oats). ③
+ Roll out the dough on a lightly floured surface to about 3mm thick and use a 5cm round cutter to cut out the biscuits. Re-roll the trimmings as necessary until you have 32 biscuits. (Prick the biscuits all over with a fork or toothpick for decorative purposes, if you don't plan to cover them with chocolate.) Place the biscuits on the lined baking sheets spaced evenly apart.
+ One tray at a time, bake the digestives for 14–18 minutes until slightly puffed up and deep golden brown.
+ Allow to cool on the baking sheets for 5–10 minutes, then transfer to a wire rack to cool completely.
+ If using the melted chocolate, dip in the cooled biscuits or dollop about ½ tablespoon of chocolate on top of each one, spreading it out into an even layer. Allow to thicken slightly at room temperature and create a wavy pattern with a small fork. Allow to set at room temperature for about 30 minutes.

STORAGE
1–2 weeks in a closed container at room temperature.

PEANUT BUTTER + JAM
THUMBPRINT COOKIES

MAKES 30
PREP TIME 45 mins
BAKE TIME 2 x 16 mins

130g natural, unsweetened, smooth peanut butter
100g unsalted butter, softened
100g light brown soft sugar
2 eggs, separated
2 tablespoons whole milk, at room temperature
½ teaspoon vanilla bean paste
200g gluten-free flour blend
½ teaspoon xanthan gum
½ teaspoon salt
150g unsalted, raw peanuts, toasted and chopped
5–6 tablespoons raspberry or strawberry jam

NOTES

① *Using a wooden spoon or spatula (rather than beating with a whisk or creaming) ensures you don't incorporate too much air into the mixture, which would result in a crumbly cookie that doesn't hold its shape well.*

② *The egg whites add moisture and the peanuts slow down the rate of heat penetration into the cookies. You can make them without rolling them in egg white and chopped peanuts, if you like, but they'll need a shorter baking time: reduce it in the next step to 11–13 minutes.*

It doesn't matter whether or not you like PBJ sandwiches, I guarantee you'll love these thumbprint cookies, with their slightly chewy interior, a crunchy crust of toasted peanuts and the jam centre that adds a pop of colour and sweet tartness. While the flavour of these cookies is spot on, it's their texture where they really shine. To make the cookies even prettier, I prefer to fill mine with jam after baking. Jam tends to spread and bubble during cooking, which isn't quite as eye-catching as a pretty dollop spooned into cooled cookies.

+ Adjust the oven shelf to the middle position, preheat the oven to 180°C and line two baking sheets with baking paper.
+ In a bowl, use a wooden spoon or spatula to mix together the peanut butter, butter and sugar until smooth. ①
+ Add the egg yolks, milk and vanilla paste, and mix until evenly incorporated.
+ Sift in the gluten-free flour, xanthan gum and salt. Mix with the spoon or spatula until you get a smooth, soft (but not sticky) dough that can be shaped into balls without cracking or crumbling.
+ Lightly whisk the egg whites in a small bowl. Place the chopped peanuts in a separate small bowl.
+ Roll about 1 tablespoon of cookie dough into a ball and roll it first in the egg whites and then in the chopped peanuts. Transfer the peanut-covered cookie ball to a lined baking sheet. Use your thumb or a ½-teaspoon measuring spoon to create a well in the centre. Repeat with the remaining dough to give 30 cookies, placing them about 2.5cm apart on the baking sheets. ②
+ One baking sheet at a time, bake the cookies for about 16–18 minutes until light golden brown. Remove from the oven and immediately use the ½-teaspoon measuring spoon to deepen the well, which might have closed up during baking.
+ Transfer the cookies to a wire rack to cool, then fill each cookie with about ½ teaspoon of raspberry or strawberry jam.

STORAGE
1 week in a closed container at room temperature.

TAHINI COOKIES

MAKES 14
PREP TIME 30 mins
CHILL TIME 15 mins
BAKE TIME 2 x 8 mins

100g tahini paste
95g maple syrup or runny honey
½ teaspoon vanilla bean paste
¼ teaspoon bicarbonate of soda
¼ teaspoon salt
100g almond flour
3 tablespoons white sesame seeds
2 tablespoons black sesame seeds

Crispy around the edges, chewy in the middle and delicious all over – it's virtually impossible not to fall in love with these tahini cookies. As their name suggests, the star of the show is the sesame paste that you may know as the crucial component of hummus, but it also works wonders in sweet treats. In this recipe, tahini gives an incredible depth of flavour, and a coating of sesame seeds makes the cookies as pretty as they are delicious.

+ Line two baking sheets with baking paper.
+ In a bowl, mix together all the ingredients except the seeds until they come together into a smooth dough.
+ In a separate small bowl, mix together the white and black sesame seeds to combine.
+ Shape 1 tablespoon of the cookie dough into a ball, roll it in the sesame-seed mixture and place it on a lined baking sheet.
+ Using the flat bottom of a glass or measuring cup, gently press down on the cookie ball until it's about 7–8mm thick.
+ Repeat with the remaining dough to give 14 cookies, then refrigerate for about 15 minutes. While the cookies are chilling, adjust the oven shelf to the middle position and preheat the oven to 180°C.
+ One baking sheet at a time, bake the cookies for about 8 minutes or until slightly spread out and light golden brown on top.
+ Immediately out of the oven, the cookies will be very soft. Allow them to cool on the baking sheets for about 10 minutes, then transfer to a wire rack to cool completely.

STORAGE
1 week in a closed container at room temperature.

PUT A TWIST ON IT
Chocolate tahini cookies – Add 15g Dutch processed cocoa powder to the cookie dough.

PEANUT BUTTER SANDWICH COOKIES

MAKES 12–13
PREP TIME 45 mins
BAKE TIME 2 x 12 mins
CHILL TIME 20 mins

Imagine everything that is good and delicious about peanut butter cups, only in cookie form. These peanut butter cookies walk the line between soft and chewy, and they simply melt in your mouth. In between, there is a luxurious, peanut butter ganache that will set off fireworks on your taste buds. And when you experience the cookies and the ganache together... peanut butter heaven.

PEANUT BUTTER COOKIES
130g natural, unsweetened, smooth peanut butter ①
100g unsalted butter, softened ②
100g light brown soft sugar
1 egg, at room temperature
190g gluten-free flour blend
½ teaspoon xanthan gum
½ teaspoon salt

PEANUT BUTTER GANACHE
150g dark chocolate (about 60% cocoa solids), chopped
150g natural, unsweetened, smooth peanut butter

NOTES
① *Using natural, unsweetened, smooth peanut butter allows you to control the amount of sugar in the recipe (and I prefer the flavour), but you can use a processed peanut butter, if you prefer. In this case, increase the flour to 210–215g to account for the palm oil, and therefore lower fibre and higher fat content.*

② *Combining butter and peanut butter allows the peanut butter flavour to shine through, while keeping the cookie dough malleable enough to prevent cracking when rolling and making the criss-cross pattern.*

PEANUT BUTTER COOKIES
+ Adjust the oven shelf to the middle position, preheat the oven to 180°C and line two baking sheets with baking paper.
+ In a bowl, use a wooden spoon or spatula to mix together the peanut butter, butter and sugar until smooth. Add the egg and combine.
+ Sift in the gluten-free flour blend, xanthan gum and salt. Mix with the spoon or spatula until the dough comes together in a ball. (It shouldn't be sticky to the touch.)
+ Tear off about 1 tablespoon of the cookie dough and shape it into a ball. Place it on a lined baking sheet, then flatten it slightly with your fingers.
+ Use the tines of a fork to create a criss-cross pattern on the top of the cookie. Repeat with the remaining cookie dough, to give 24–26 cookies, spacing them evenly over the baking sheets.
+ One baking sheet at a time, bake the cookies for about 12–15 minutes, using less time for softer, chewier cookies and more time for crisper cookies with a snap.
+ Allow to cool on the baking sheets for 10 minutes, then transfer to a wire rack to cool completely.

PEANUT BUTTER GANACHE
+ In a heatproof bowl over a pan of simmering water, melt the dark chocolate and peanut butter together until smooth.
+ Remove from the heat and allow to cool at room temperature, stirring occasionally, then refrigerate for about 20–30 minutes until the ganache reaches a pipeable consistency.
+ Pipe about 1 tablespoon of the ganache on to the flat (under)side of one cookie and sandwich it with another, flat sides facing. Repeat with the remaining cookies to make 12–13 sandwich cookies.

STORAGE
1 week in a closed container in a cool, dry place

CHOCOLATE CRINKLE COOKIES

MAKES 20
PREP TIME 30 mins
CHILL TIME 1 hour
BAKE TIME 2 x 10 mins

150g caster sugar
50g vegetable, sunflower or other
   neutral-tasting oil ①
2 eggs, at room temperature
½ teaspoon vanilla bean paste
120g gluten-free flour blend
60g Dutch processed cocoa
   powder
½ teaspoon xanthan gum
1 teaspoon baking powder
¼ teaspoon salt
60g icing sugar

NOTE

① *Although I normally swear
   by using butter in my cookies,
   chocolate crinkle cookies are
   much better made with oil –
   they're lighter and moister
   (and they stay that way longer).*

What's not to love about chocolate crinkle cookies? They're chocolatey, fudgy and covered in icing sugar – plus, they're incredibly fun to make. It's pretty much impossible to resist the temptation to peek into the oven, every minute or so, to see them transform from little icing-sugar snowballs into crackly, black-and-white beauties.

+ In a bowl, whisk together the sugar and oil – the mixture will have the texture of wet sand.
+ Add the eggs and vanilla paste, and whisk until smooth.
+ Sift in the gluten-free flour blend, cocoa powder, xanthan gum, baking powder and salt. Fold together with a spoon or spatula to a smooth cookie batter – it will resemble brownie batter, flowing off the spoon or spatula.
+ Cover the bowl with cling film and refrigerate for at least 1 hour to firm up.
+ Adjust the oven shelf to the middle position, preheat the oven to 180°C and line two baking sheets with baking paper.
+ Check the cookie dough – if it still feels too soft and sticky, leave it in the fridge for a further 30 minutes.
+ When the dough is firm and easy to handle, spoon out about 1 heaped tablespoon and shape it into a ball. Roll it in icing sugar to coat heavily and place it on a lined baking sheet. Repeat with the remaining dough to give 20 cookies altogether, spacing them at least 2.5cm apart.
+ One baking sheet at a time, bake the cookies for about 10 minutes or until they have cracked on top, but are still soft when you gently press down on them.
+ Allow to cool on the baking sheets for 10 minutes, then transfer to a wire rack to cool completely.

STORAGE

1 week in a closed container at room temperature.

PUT A TWIST ON IT

To ramp up the chocolate flavour, add chocolate chips or melted chocolate to the cookie batter. Or, try one of the following flavour variations (with or without the extra chocolate):

Mexican hot chocolate crinkle cookies – Add a pinch of cayenne pepper and cinnamon.

Orange + chocolate crinkle cookies – Add orange zest.

GINGERBREAD COOKIES

MAKES 42
PREP TIME 1 hour
BAKE TIME 2 x 6 mins

Christmas simply wouldn't be Christmas without a few batches of gingerbread making your kitchen smell like sugar and spice (and all things nice). It's incredibly important that gingerbread cookies keep their shape (so that you can decorate them), while also being chewy with delicious, crisp edges. These cookies are all that, without even needing to chill the dough (unless your kitchen is especially warm).

GINGERBREAD COOKIES

75g unsalted butter, softened
75g light brown soft sugar
100g molasses
1 egg yolk, at room temperature ①
260g gluten-free flour blend
¼ teaspoon xanthan gum
¼ teaspoon bicarbonate of soda ②
1 teaspoon ground cinnamon
1 teaspoon ground ginger
¼ teaspoon allspice
¼ teaspoon ground cloves
¼ teaspoon salt

ROYAL ICING

10g egg-white powder
7 tablespoons warm water
200g icing sugar, sifted
2 teaspoons lemon juice
food colouring (optional)

NOTES

① *Using a yolk rather than a whole egg gives the cookies a slightly chewier texture, and reduces their rise and spread in the oven.*

② *This amount of bicarbonate of soda is small enough to give the cookies very slightly rounded edges without making them spread out or lose their shape.*

③ *Thinner cookies hold their shape better. Cookies thicker than 3mm puff up and spread more, giving a softer, more rounded look.*

GINGERBREAD COOKIES

+ Adjust the oven shelf to the middle position, preheat the oven to 180°C and line two baking sheets with baking paper.
+ Using a stand mixer with the paddle attachment or a hand mixer fitted with the double beaters, cream the butter and sugar together until pale and fluffy. Add the molasses and egg yolk, and mix until incorporated.
+ Sift in the remaining cookie ingredients, then mix well until the dough starts coming together in a ball.
+ Turn out the dough on to a lightly floured surface and knead briefly until it's completely smooth, fairly firm and not sticky. Roll out to 2–3mm thick. ③ If it's too soft to cut, chill for about 15 minutes. (Slide an offset spatula under the dough to loosen any stuck-on patches.)
+ Cut out the cookies using a cookie cutter of choice, re-rolling the trimmings as necessary (a 6cm cutter gives about 42 cookies).
+ Place the cut cookies on the lined baking sheets, about 1cm apart.
+ One baking sheet at a time, bake for 6–8 minutes for cookies with a bit of chew; or up to 10 minutes for crisper cookies (for example, to use as Christmas-tree decorations).
+ Allow the cookies to cool on the baking sheets for about 5–10 minutes to firm up a little, then transfer to a wire rack to cool completely.

ROYAL ICING

+ In a bowl, rehydrate the egg-white powder with 2 tablespoons of the warm water, whisking until fully dissolved and slightly frothy.
+ Add the icing sugar, lemon juice and remaining warm water and, using a stand mixer with the whisk attachment or a hand mixer fitted with the double beaters, whisk the mixture until soft peaks form.
+ Adjust the consistency of the icing: add more water (1 teaspoon at a time) or icing sugar (1 tablespoon at a time) to make it looser or thicker, as you prefer. If you need coloured icing, add food colouring now too.
+ Transfer the royal icing to a piping bag fitted with a small round nozzle and decorate the cooled gingerbread cookies.
+ Leave the cookies at room temperature for 2–3 hours for the icing to set.

STORAGE

1–2 weeks in a closed container at room temperature. (Any unused royal icing will keep in an airtight container for 1 week at room temperature.)

ROSEMARY CRACKERS

MAKES 22–24
PREP TIME 30 mins
BAKE TIME 2 x 10 mins

125g gluten-free flour blend
½ teaspoon xanthan gum
1½ teaspoons baking powder
½ tablespoon caster sugar
½ teaspoon salt
2 rosemary sprigs, leaves picked
  and chopped
45g unsalted butter, cubed and
  chilled
1 tablespoon sunflower or other
  neutral-tasting oil (or olive oil
  to add flavour)
30g cold water
1½ tablespoons melted unsalted
  butter
flaky sea salt, for sprinkling

NOTES

① *If you don't have a food
  processor, rub the butter into
  the dry ingredients with your
  fingers, and use a fork to bring
  the dough together after you've
  added the oil and water.*

② *Brushing the hot crackers with
  butter gives them a more delicate
  texture and a richer flavour.*

Forget store-bought crackers – these are almost dangerously delicious in all of their salty, buttery, rosemary-scented glory. They strike the perfect balance between being crisp and sturdy enough for dipping, while also being crumbly and buttery enough to have an incredibly rich mouthfeel. Plus, they're easy to make, with the food processor doing most of the work for you.

+   Adjust the oven shelf to the middle position, preheat the oven to 200°C and line two baking sheets with baking paper.
+   Add the gluten-free flour blend, xanthan gum, baking powder, sugar, salt and chopped rosemary to the bowl of a food processor, and blend briefly to mix thoroughly. ①
+   Add the cold butter and process until you get a coarse-sand texture, then add the oil and process until evenly distributed.
+   Add the cold water, 1 tablespoon at a time, processing after each addition, until the dough starts coming together in a ball. Transfer the dough to the work surface and give it a gentle knead to make it smooth.
+   Roll out the dough on a lightly floured work surface to about 1–2mm thick, and use a 6cm round cutter to cut out the crackers, re-rolling the trimmings as necessary, until you have 22–24 crackers. Transfer the crackers to the lined baking sheets and prick them all over with a fork or toothpick.
+   One baking sheet at a time, bake the crackers for about 10–12 minutes or until slightly puffed up and light golden brown. Remove from the oven and immediately brush with melted butter and sprinkle with flaky sea salt. ② Allow to cool on the baking sheets for about 5 minutes, then transfer to a wire rack to cool completely.

STORAGE
1 week in a closed container at room temperature.

PUT A TWIST ON IT
Spiced rosemary crackers – Add a pinch of cayenne pepper and smoked paprika for a spicier version.
Mediterranean rosemary crackers – Add some garlic powder and a mixture of herbs, such as rosemary, oregano and thyme.

PEANUT-BUTTER-STUFFED CHOCOLATE CHUNK SKILLET COOKIE

SERVES 8
PREP TIME 45 mins
COOK TIME 6 mins
BAKE TIME 30 mins

There's so much to love about this skillet cookie: the brown butter, which gives it a deep, nutty flavour; the abundance of chocolate, especially when it's still warm and melted and luscious; the crisp, caramelised edges; the gooey centre; the dollops of peanut butter, making every mouthful even more exciting; and, finally, the opportunity to spoon it out in glorious chunks and serve it with a few generous scoops of ice cream.

180g unsalted butter, plus extra
  for greasing
110g caster sugar
180g light brown soft sugar
1 egg, at room temperature
2 egg yolks, at room
  temperature ①
½ teaspoon vanilla bean paste
315g gluten-free flour blend
¾ teaspoon xanthan gum
1½ teaspoons baking powder
½ teaspoon salt
150g dark chocolate (60–70%
  cocoa solids), chopped
125g natural, unsweetened,
  smooth peanut butter
flaky sea salt, for sprinkling
vanilla ice cream, to serve
  (optional)

NOTES

① *The egg yolks make the skillet cookie extra-rich and fudgy or gooey.*

② *If you don't have a cast-iron skillet, you can use a 25cm round cake tin instead.*

+ First, brown the butter. Add it to a saucepan (preferably one of a light colour so you can see the butter changing colour) and cook it over a medium heat, stirring frequently, for about 5–6 minutes until melted. It will then start bubbling and foaming, and turn amber. You should see specks of a deep brown colour on the bottom of the saucepan – those are the caramelised milk solids. This should take about 6–8 minutes in total. Remove the butter from the heat and set aside to cool until warm.
+ Adjust the oven shelf to the middle position, preheat the oven to 180°C and generously butter a 25cm cast-iron skillet. ②
+ Using a stand mixer with the paddle attachment or a hand mixer fitted with the double beaters, cream the brown butter and both sugars together until pale and fluffy.
+ Add the egg, egg yolks and vanilla paste, and mix to fully incorporate.
+ In a separate bowl, sift together the gluten-free flour blend, xanthan gum, baking powder and salt. Add the dry ingredients to the wet in two batches, mixing well after each addition. Stir in 125g of the chopped dark chocolate.
+ Press half the cookie dough into the buttered skillet. Drop spoonfuls of peanut butter on top of it, then crumble the remaining cookie dough on top. Scatter the reserved chopped chocolate over it.
+ Bake for 30–40 minutes (30 minutes for a gooey cookie; 40 for fudgy), until the top layer of cookie dough has spread out and is golden brown and the centre is still slightly underbaked. If the top starts browning too quickly, cover with foil (shiny side up) and bake until done.
+ Allow to cool for 10–15 minutes, then sprinkle with flaky salt and serve warm, with a scoop of vanilla ice cream, if you wish.

PUT A TWIST ON IT
Choco-hazelnut-stuffed chocolate chunk skillet cookie – Simply substitute spoonfuls of chocolate hazelnut spread for the peanut butter.
Gingerbread chocolate chunk skillet cookie – Omit the peanut butter and instead add 2 teaspoons each of ground cinnamon and ground ginger, ½ teaspoon allspice and 3–4 tablespoons molasses to the cookie dough.

SALTED-CARAMEL-STUFFED CHOCOLATE CHIP COOKIE BARS

MAKES 12–16
PREP TIME 1 hour
COOK TIME 15 mins
BAKE TIME 40 mins

CHOCOLATE CHIP COOKIE DOUGH

150g unsalted butter, melted and cooled

75g caster sugar

150g light brown soft sugar

1 egg, at room temperature

2 egg yolks, at room temperature ①

½ teaspoon vanilla bean paste

315g gluten-free flour blend

¾ teaspoon xanthan gum

1½ teaspoons baking powder

½ teaspoon salt

150g dark chocolate chips

CARAMEL LAYER

180g double cream

30g unsalted butter

80g golden syrup

100g caster sugar

¼–½ teaspoon salt, to taste

The idea behind this recipe is simple, yet incredibly effective: take your average chocolate chip cookie, supersize it and stuff it with salted caramel that oozes out of the middle. (Sounds amazing, doesn't it?) Of course, it's not quite as straightforward as that – to get just the right combination of flavours and textures, you need to increase the amount of flour in a standard recipe for chocolate chip cookies (lest you end up with a greasy mess) and the caramel layer is a careful balance of cream, sugar and golden syrup, cooked to just the right temperature. But get those things right (follow the recipe and you will), and you'll end up with bars that are crispy around the edges, gooey in the middle and have luscious caramel oozing out of them, beckoning you back for a second (and third, and just-one-more) piece.

CHOCOLATE CHIP COOKIE DOUGH

+ Line a 20cm square baking tin with a piece of baking paper cut large enough to overhang the sides (to help you remove the baked bars).

+ Using a stand mixer with the paddle attachment or a hand mixer fitted with the double beaters, cream the butter and both sugars together until pale and fluffy.

+ Add the egg, egg yolks and vanilla paste, and mix to fully incorporate.

+ Sift together the gluten-free flour blend, xanthan gum, baking powder and salt. Add the dry ingredients to the wet in two batches, mixing well after each addition. Stir in the chocolate chips.

+ Press half the cookie dough (about 470g) into the bottom of the lined baking tin, creating a 5mm raised rim around the edge (to prevent the caramel from leaking out too much during baking).

+ Refrigerate the cookie dough layer in the baking tin and the remaining cookie dough until needed.

CARAMEL LAYER + ASSEMBLING THE BARS

+ Adjust the oven shelf to the middle position and preheat the oven to 180°C.
+ In a saucepan, bring the double cream and 15g of the butter to the boil. Remove from the heat and set aside until needed, making sure the mixture stays warm.
+ In a clean saucepan, combine the golden syrup and sugar, and cook over a medium–high heat, stirring frequently, until it reaches 145°C. Remove from the heat, add the cream mixture and combine, then heat again to 112–116°C, stirring continuously. ② Immediately remove from the heat again and stir in the salt and the remaining butter.
+ Stirring occasionally, cool the caramel to 80°C – at this temperature, it's unlikely to melt the bottom cookie dough layer while still being pourable and spreadable. Pour this over the bottom cookie dough layer and spread it out, making sure it doesn't go over the raised cookie-dough edge.
+ Crumble the other half of the cookie dough evenly over the caramel layer, completely covering the caramel.
+ Bake for 40–45 minutes until the top layer has spread out and is golden brown. If the top starts browning too quickly, cover with foil (shiny side up) and bake until done.
+ Allow to cool until warm, remove from the tin and cut into bars to serve.

STORAGE

3–4 days in a closed container at room temperature. If you want the caramel centre to be extra-soft and oozy, reheat individual bars in the microwave for 5–10 seconds before serving.

PUT A TWIST ON IT

Chocolate + hazelnut-stuffed chocolate chip cookie bars – Stuff the cookie bars with chocolate hazelnut spread, instead of salted caramel, and add chopped, toasted hazelnuts to the cookie dough.

NOTES

① *The egg yolks make the cookie bars extra-rich and fudgy or gooey.*

② *Between 112°C and 116°C, the caramel is in the so-called 'softball' stage. This is the optimal temperature range to give a caramel layer that, despite being heat-treated again during baking, is luxuriously soft even once the finished cookie bars have cooled completely. To test for the softball stage, drop a small amount of the caramel into a glass of cold water and then scoop it out – it should hold its shape but be very soft when squeezed. (You don't need to do this if you're using a sugar thermometer.)*

PHOTOGRAPHY OVERLEAF ⟶

BLUEBERRY PIE CRUMBLE BARS

MAKES 12–16
PREP TIME 45 mins
CHILL TIME 20 mins
COOK TIME 10 mins
BAKE TIME 40 mins

You might think that pre-cooking the blueberry filling and reducing the juices is a bit overkill – but the final result is worth that extra effort a thousand times over. The blueberry layer is an explosion of juicy flavour, without making the surrounding cookie dough soggy. Instead, the cookie layers are buttery and crumbly, but still hold their shape just enough to be able to cut them into pretty little slices.

CRUMBLE COOKIE DOUGH

200g unsalted butter, softened
150g light brown soft sugar
1 egg, at room temperature
½ teaspoon vanilla bean paste
360g gluten-free flour blend
1 teaspoon xanthan gum
1 teaspoon baking powder
¼ teaspoon salt
25g ground almonds

BLUEBERRY FILLING

500g fresh or frozen blueberries
125g caster sugar
3 tablespoons lemon juice

NOTES

① Chilling the other half of the cookie dough sets the butter and makes it easier to crumble over the filling, as well as helping it hold its shape during baking, just enough to give a beautifully textured top.

② Pre-cooking the blueberry filling and reducing the juices concentrates their flavour and reduces the moisture content to prevent the bars going soggy during baking.

CRUMBLE COOKIE DOUGH

+ Line a 20cm square baking tin with a piece of baking paper cut large enough to overhang the sides (to help you remove the baked bars).
+ Using a stand mixer with the paddle attachment or a hand mixer fitted with the double beaters, cream the butter and sugar until pale and fluffy. Add the egg and vanilla paste, and mix to incorporate.
+ Sift in the gluten-free flour blend, xanthan gum, baking powder and salt and mix until you get a smooth, soft (but not sticky) dough.
+ Press half the cookie dough (about 380g) into the bottom of the lined baking tin and create a smooth, even layer. Set aside until needed.
+ Mix the ground almonds into the other half of the cookie dough and refrigerate for at least 20–30 minutes. ①

BLUEBERRY FILLING + ASSEMBLING THE BARS

+ Adjust the oven shelf to the middle position. Preheat the oven to 180°C.
+ Combine the blueberries, sugar and lemon juice in a large saucepan and cook over a medium–high heat until the blueberries have softened and released their juices. Drain the blueberries through a sieve, collecting the juices in a bowl or jug. Tip the blueberries into a bowl and set aside.
+ Return the juices to the saucepan and reduce them over a medium–high heat until they have thickened and become viscous, but aren't quite at the jam stage yet (about 5 minutes). Remove from the heat and allow to cool completely, then mix in the whole blueberries. ②
+ Pour the blueberry filling over the cookie layer and crumble over the chilled cookie dough, completely covering the filling.
+ Bake for about 40–45 minutes or until the top layer of dough has spread out slightly and is golden brown. If it starts browning too quickly, cover with foil (shiny side up) and continue baking until done.
+ Allow to cool completely in the tin, then remove and cut into slices.

STORAGE

3–4 days in a closed container at room temperature.

PUT A TWIST ON IT

Try these bars with other berries (blackberries or raspberries, say), or with cherries or apricots. If you're short on time, use jam as the filling.

S'MORES BARS

MAKES 16
PREP TIME 1 hour 15 mins
BAKE TIME 30 mins
CHILL TIME 3 hours

Let me tell you about these s'mores bars. There's the crumbly, buttery cookie base (tasting like a digestive biscuit – after all, we are in s'mores territory). On top is a layer of luxurious chocolate ganache. And finally, the part that makes these bars extra special: homemade marshmallows, piped in pretty little mounds of fluffy, sugary deliciousness and toasted to golden-brown perfection. Don't worry – making marshmallow is actually really easy and, I promise you, the results are so worth it.

COOKIE BASE

60g gluten-free flour blend
60g brown rice flour
60g gluten-free oats, coarsely ground
50g light brown soft sugar
¼ teaspoon xanthan gum
¼ teaspoon salt
145g unsalted butter, cubed and chilled

CHOCOLATE GANACHE

255g dark chocolate (about 60% cocoa solids), chopped
350g double cream

MARSHMALLOW TOPPING

8g gelatine powder
140g water
200g caster sugar
½ teaspoon vanilla bean paste

COOKIE BASE

+ Adjust the oven shelf to the middle position, preheat the oven to 180°C and line a 20cm square baking tin with a piece of baking paper cut large enough to overhang the sides (to help you remove the baked bars).
+ Add the gluten-free flour blend, brown rice flour, ground oats, sugar, xanthan gum and salt to the bowl of a food processor and pulse a few times to combine. ①
+ Add the butter and pulse until fully incorporated and you get a mixture of smaller breadcrumb-like pieces and larger clumps. The mixture should hold its shape if you squeeze it together.
+ Transfer to the lined baking tin and use your fingers or the base of a glass or measuring cup to press it into a compact, even layer.
+ Bake for about 30 minutes or until it's evenly deep golden brown. Remove from the oven and set aside to cool.

CHOCOLATE GANACHE

+ Place the chopped dark chocolate into a heatproof bowl.
+ In a saucepan, bring the double cream just to the boil, then pour it over the chocolate. Allow the mixture to stand for 4–5 minutes, then stir to a smooth, glossy ganache.
+ Pour the ganache over the cooled cookie base, gently shaking the tin to settle it into an even layer. Refrigerate for 2–3 hours, or preferably overnight, for the ganache to firm up and set.

MARSHMALLOW TOPPING ②

+ In a small bowl, combine the gelatine powder and 80g of the water. Stir and allow to sit for 10 minutes, then add the thickened mixture to the bowl of a stand mixer (or a large bowl if using a hand mixer).
+ In a saucepan, combine the caster sugar and the remaining water, and cook over a medium–high heat, stirring occasionally, until the syrup reaches 110–115°C.
+ Add the hot syrup to the gelatine, mixing briefly to combine.
+ Using a stand mixer with the whisk attachment or a hand mixer fitted with the double beaters, whisk the mixture on high speed for 5–8 minutes until cooled slightly, tripled in volume, soft peaks form and the marshmallow holds its shape. Use the marshmallow as soon as it's ready, as it starts setting fairly quickly.
+ Transfer the marshmallow to a piping bag fitted with a large open star nozzle and pipe it on top of the set ganache layer.
+ Return the bars to the fridge for at least 1 hour to set the marshmallow.
+ To serve, use a hot knife to cut into slices, wiping the knife clean between cuts. Allow the bars to come up to room temperature for 5–10 minutes, then, using a kitchen blow torch, toast the tops to brown the marshmallow slightly.

STORAGE

The untoasted bars will keep for 3–4 days in the fridge, then toast just before serving.

NOTES

① *This combination of gluten-free flours is the same as for the digestive biscuits (see page 162) – the flavour and texture is therefore just like the regular biscuits that you might use to make s'mores.*

② *While you could definitely make a Swiss meringue topping (as for the toasted marshmallow brownies on page 142), piping homemade marshmallows on top of the ganache layer gives a more authentic s'mores experience.*

PHOTOGRAPHY OVERLEAF ⟶

LEMON BARS

MAKES 16
PREP TIME 30 mins
COOK TIME 5 mins
BAKE TIME 42 mins
CHILL TIME 3 hours

This recipe pre-cooks the lemon filling on the hob, shortening the oven time and ensuring that it is silky smooth all the way through – no overbaked edges or underbaked centre in sight. At the same time, the gluten-free flour blend:butter ratio in the shortbread guarantees a crust that is easy to slice, is just crumbly enough and melts in your mouth. And if you're like me and don't like your lemon bars ultra-sour, you can tweak the lemon juice:water ratio to suit your taste (as long as the total liquid weight remains 230g).

SHORTBREAD BASE

175g gluten-free flour blend
50g caster sugar
¼ teaspoon xanthan gum
¼ teaspoon salt
145g unsalted butter, cubed
  and chilled

LEMON FILLING

4 eggs
5 egg yolks ①
300g caster sugar
½ teaspoon salt
160g lemon juice
70g water
65g unsalted butter
zest of 2 unwaxed lemons
  (optional)

NOTES

① *Using a larger number of egg yolks to thicken and set the filling is preferable to using cornflour or gluten-free flour blend, which can make the filling murky and destroy the silky-smooth texture.*

② *This short period in the oven helps to set the filling and smooths out the top, giving very neat bars that are easy to slice.*

SHORTBREAD BASE

+ Adjust the oven shelf to the middle position, preheat the oven to 180°C and line a 20cm square baking tin with a piece of baking paper cut large enough to overhang the sides (to help you remove the baked bars).
+ Add the gluten-free flour blend, sugar, xanthan gum and salt to the bowl of a food processor and pulse a few times to combine.
+ Add the butter and pulse until fully incorporated and you get a mixture of smaller breadcrumb-like pieces and larger clumps. The mixture should hold its shape if you squeeze it together.
+ Transfer to the lined baking tin and use your fingers or the base of a glass or measuring cup to press it into a compact, even layer.
+ Bake for about 30 minutes or until it's evenly golden brown. Remove from the oven and set aside to cool slightly while you make the lemon filling.

LEMON FILLING

+ Reduce the oven temperature to 160°C.
+ In a non-reactive saucepan, whisk the eggs, egg yolks, sugar and salt until smooth. Add the lemon juice and water, and whisk until combined.
+ Cook the mixture over a medium–high heat, stirring continuously, for about 5–10 minutes until noticeably thickened to a loose, pudding-like consistency (about 76–78°C in temperature).
+ Remove from the heat, add the butter, and the lemon zest (if using), and mix well until the butter is melted and the mixture is smooth and homogeneous.
+ Pour the lemon filling over the warm shortbread base, smooth out the top and bake for 12–15 minutes or until a 2–3cm border is set but the centre is still wobbly when you shake the tin. ②
+ Cool in the tin for 1 hour at room temperature, then refrigerate for at least 3 hours, or preferably overnight. Once chilled and set, remove from the baking tin, dust with icing sugar and cut into 16 squares.

STORAGE
3–4 days in a closed container in the fridge.

PIES, TARTS + PASTRIES

You should know… some of the recipes in this chapter will change your life. Well, your food life at least. It might be just me, but I swear I get a bit teary-eyed when I take the very first bite of the caramelised onion and cherry tomato tartlets (see page 235), and if there's the strawberries and cream tart (see page 240) for dessert, I'm a goner.

I could write essays filled with superlatives about each and every one of the recipes on the following few pages (admittedly, I am somewhat biased), but it's the gluten-free rough puff pastry that really gets me. For the longest time, my version of gluten-free rough puff pastry was just flaky pie crust with a bit of extra lamination. Don't get me wrong, it wasn't bad – it was wonderfully flaky and crisp, but I wanted the real deal, or as close as I could get.

Enter grated frozen butter. Now, the idea of making a smooth dough, lower in butter, and then incorporating grated frozen butter into it through a series of folds isn't exactly a new idea in the world of pastry. In fact, it's been around for quite a while. But, for some reason, I'd never thought to apply it to gluten-free rough puff pastry.

The moment I did (and after a bit of fiddling around with the ingredient ratios and the number of folds), it was as though a whole new world of pastry opened up before me. The results this new recipe gave – the rise and the flake and the delicate, well-defined layers of crisp pastry that made that characteristic 'crunch' sound as you bit into them – it was everything that makes puff pastry great, only easier and quicker to make… and gluten-free. (And when I say quicker: the laminating process takes 20 minutes, tops.)

This discovery resulted in some truly wonderful new recipes, and led me to simplify other basic pastry recipes, too. For example, both the plain and the chocolate sweet shortcrust pastry are easy recipes that you can prepare in about 15 minutes, before chilling them in the fridge for a bit. All this is to say: pastry isn't scary, even if it's gluten-free.

THE TYPES OF PIE, TART + PASTRY IN THIS CHAPTER

All the recipes in this chapter use one of four fundamental pastry recipes: **flaky all-butter pie crust** (see page 200), **rough puff pastry** (see page 203), **plain sweet shortcrust pastry** (see page 204) and **chocolate sweet shortcrust pastry** (see page 207). As all other recipes begin with making one of these types of pastry, and because repeating the same ingredients and instructions again and again is: (a) tedious and (b) a rather poor use of space (so many more interesting things to say!), I've written up these recipes separately and referred to them when relevant throughout the chapter. They are scaled appropriately, so that you'll need to make a half or a full quantity, according to the recipe – no need for complicated maths, just divide by two where necessary.

There's another reason for looking at the four fundamental recipes separately. With pastry, it's often the case that you have a particular recipe in mind that you really want to make, and the only component you need to convert to gluten-free is the pastry. Instead of having to look at a specific recipe for this or that pie, tart or other dessert, and extracting only the information pertaining to the pastry, I've already done this for you. (Hopefully, these pros far outweigh the solitary con of having to flick through a few additional pages when making a single recipe – something I've tried to avoid elsewhere in the book.)

Beyond these fundamentals, the recipes span a wide range of both sweet and savoury deliciousness, chosen not only because they're amazingly beautiful taste explosions (and they are), but because pretty much every one of them introduces an interesting technique, method of preparation or a titbit of science I really think you should know.

As this chapter could get confusing very quickly (what with four types of pastry, some in both sweet and savoury versions), the recipes are organised according to pastry type, starting with those using flaky pie crust and ending with chocolate shortcrust. Then, within each group we start with sweet and end with savoury (where applicable).

WHEAT-BASED VS GLUTEN-FREE PASTRY

In addition to the fact that they're delicious, shortcrust pastry, pie crust and rough puff pastry all have the highly important job of keeping your chosen filling contained so that it doesn't leak all over the place – that is, their role is to a large extent structural. Knowing that, and considering the facts that gluten provides structure and elasticity to wheat-based bakes and that pastry contains a pretty large amount of flour (see Table 3, page 34), it's no surprise that the absence of gluten is felt more strongly in this chapter than, for example, in the brownies.

Thankfully, the problem is easy to solve. All we have to do is add some xanthan gum to the dry ingredients in an appropriate quantity. For example, the mealy, slightly crumbly shortcrust pastry requires less xanthan gum than flaky pie crust and rough puff pastry.

With shortcrust pastry, that's pretty much the extent of the difference between wheat and gluten-free versions. For flaky pie crust and rough puff, there's a further thing to consider – the amount of water you add.

As I've said before (see page 15), gluten-free flours have a higher water-absorption capacity than wheat flour, which basically means that you'll need to add more water both to gluten-free flaky pie crust and to gluten-free rough puff than to their wheat-based equivalents.

But if you do these two simple tweaks – add xanthan gum and increase hydration – you'll end up with pastry that handles beautifully (no tearing or cracking in sight) and bakes to golden perfection.

THE STRUCTURE OF PASTRY

To understand the preparation methods of the different pastry types, we also need to have a thorough look at their structures and how they determine the properties of the finished baked product.

Rough puff pastry (see page 203) is a handy substitute for regular puff pastry, and far quicker and easier to assemble at home than the 'real deal'. The process of lamination creates fairly well-defined layers of dough and butter – I say 'fairly well-defined' because they're nowhere near as neat and regular as those of actual puff pastry, but they definitely get the job done.

The primary characteristic of rough puff pastry is its ability to puff up when baked, resulting in an open, flaky structure. For this, lamination is crucial. As the pastry encounters the hot temperatures of the oven, the water contained in the butter layers (about 15–16% of butter by weight) rapidly evaporates and expands, and the pastry puffs up. At the same time, the butter melts away and is partially absorbed by the dough layers, leaving behind a tender, flaky pastry.

Flaky all-butter pie crust (see page 200) is essentially a less laminated, less 'puffed up' and consequently sturdier version of rough

ROUGH PUFF PASTRY

LAYER OF BUTTER

GF FLOUR PARTICLES

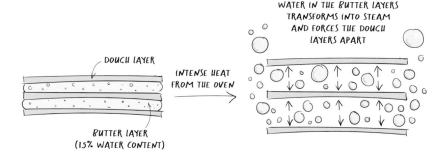

DOUGH LAYER

INTENSE HEAT FROM THE OVEN

BUTTER LAYER (15% WATER CONTENT)

WATER IN THE BUTTER LAYERS TRANSFORMS INTO STEAM AND FORCES THE DOUGH LAYERS APART

The lamination (alternate layers of butter and dough) in rough puff pastry and, to a smaller extent, flaky pie crust affects the behaviour of the pastry in the oven, causing it to puff up into distinct, flaky layers.

puff pastry. It is perfectly suited to rustic desserts such as apple pie, where the pastry needs to be robust enough to hold the filling without becoming soggy or falling to pieces when you try to cut a slice. Additionally, pastry greatly puffing up in the oven is rather unnecessary (and impossible) when it's either covered with filling or blind baked.

For this pastry, you work butter into the flour until quite large pieces remain. Through a series of several folds, you can achieve very rudimentary lamination. The purpose of this lamination is three-fold:

+   without it, the pastry would be much tougher and harder texture-wise, making for a less pleasant eating experience – instead, the pastry is perfectly flaky and delicate,

+   the additional butter–dough layers give just enough rise to the pastry to make it look gorgeous when used for pie lattices or lids,

+   with each fold, the pastry becomes noticeably smoother, easier to handle and even less prone to cracking.

As it happens, bakers often use this method to make homemade rough puff pastry – which should tell you something about how much flakiness to expect (that is: a whole lot). Still, in my opinion it's nowhere near the levels of flakiness and rise you would (and should) expect from puff pastry, so I tend to reserve it for pies, where it's simply perfect.

Sweet shortcrust pastry (see pages 204 and 207) is on the completely opposite end of the spectrum. There's no lamination, no flake and absolutely no puffing up in the oven (unless something goes wrong). Instead, the pastry is buttery and walks the delicate line between being crisp and sturdy enough to hold its shape and any fillings, and being crumbly enough to melt in your mouth.

Accordingly, the method of preparation is completely different. Instead of cold butter, you use softened butter – although we still need to mix it with dry ingredients as with the flaky pie crust. However, instead of leaving the butter in large pieces, the final texture of the flour–butter texture resembles coarse sand or breadcrumbs. Adding an egg brings the pastry together into a dough that's a joy to work with.

FLAKY PIE CRUST

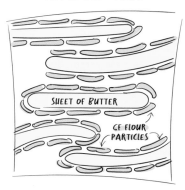

SHEET OF BUTTER

GF FLOUR PARTICLES

SHORTCRUST PASTRY

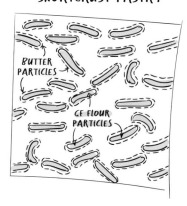

BUTTER PARTICLES

GF FLOUR PARTICLES

HOW (WHEN AND WHY) TO BLIND BAKE PASTRY

Blind baking is a central process in pie and tart making, so it's important to understand not only the how but also the when and why.

You should blind bake your pastry when:

+ **You won't bake the filling**, but rather use a pre-cooked filling (like in the lemon meringue pie on page 216) or use a filling that doesn't require cooking (like the chocolate mousse filling in the hot chocolate tart on page 244). In this case, you need to blind bake the pastry case until it's fully cooked.

+ **The filling you're adding is extra wet**, like in a pumpkin pie, or if the baking time once you add the filling needs to be relatively short so as not to overbake. In this case, you need to only partially blind bake the pastry case, as it will have to go back into the oven for a short while once you've filled it.

You shouldn't blind bake your pastry when:

+ **You're using rough puff pastry** – unless you're making millefeuille (see page 350) – because compressing the pastry rather defeats the purpose of making rough puff in the first place.

+ **The combined blind baking and baking with the filling would dry out, overly brown or even burn the pastry**. That's the case with the raspberry frangipane tart (see page 239), where blind baking would result in an unattractive dark brown, bone-dry pastry shell. That's why the filling is put directly into the raw pastry shell, and everything is baked together.

+ **You're making a double-crust pie**, where the top is either a lattice or a full pastry lid. Of course, there are ways of blind baking a double-crust pie, but they can be slightly fussy. And, in any case, if you follow my advice below on how to avoid a soggy bottom, blind baking the base of a double-crust pie is unnecessary.

Now that you know when and why you should (or shouldn't) blind bake your pastry, let's look at the technicalities of actually doing it. Starting with the pastry arranged snugly in a baking tin or pie dish of choice, with the edges trimmed and crimped if relevant, it's best that you refrigerate it briefly before proceeding. This sets and firms up the pastry, ensuring that when you prepare it for blind baking, you don't in any way damage it.

Once you've chilled the pastry case, prick it all over the base with a fork – although it will be weighed down during blind baking, pricking is an additional insurance against the pastry puffing up because there is any air or moisture trapped underneath with no way to escape. Then, you need to line the pastry case with a piece of baking paper.

Arranging a pristine, freshly cut piece of baking paper so that it fits to the size of the pastry is a bit of a nightmare. What's worse, though, is that it's a nightmare that can tear your pastry with stray sharp corners where the baking paper bends. That's why you should scrunch up the baking paper, making it soft and malleable.

Baking beans – or plain old rice

Once you've lined the pastry with the baking paper, you need to fill it with something that will weigh it down, preventing it from puffing up and shrinking away from or sliding down the sides of the pie or tart tin. You can go the fancy route and use ceramic baking beans, but I usually just use good-old rice. The individual grains of rice are smaller than individual baking beans, meaning that rice is more efficient at adapting to the shape of the pastry and the tin you're using. Pour in the rice all the way to the rim of the pastry case and press it down gently to settle into the nooks and crannies. You can't use the rice for cooking and eating afterwards, but don't throw it away – save it and re-use it in future blind-baking endeavours.

In the oven

Just like when you're baking filled pies, it's best to blind bake a pie crust on the lower-middle shelf of the oven on a preheated baking tray or sheet (the hot tray or sheet ensures that the pie base crisps up and bakes fully before the edges start getting too dark).

Begin by baking the pie crust at 200°C for 18–20 minutes until you can see the edges turning a light golden colour. Then, remove it from the oven and take out the rice or baking beans and the baking paper. If your pie crust is to have an unbaked filling, you need to bake the crust through fully on its own. To do this, once you've removed the paper and rice or beans, return the crust to the oven and bake it for a further 10–15 minutes until it looks, feels and sounds crisp. It should have a uniform golden-brown colour. The mark of a perfectly blind-baked crust is that you can easily pick it up out of the pie dish, without any sticking, bending or cracking.

There's more than just a crisp crust with a gorgeous colour that comes from the second baking, though: the extra stint in the oven dries out the outer surface of the pastry and almost seals it, so that it doesn't absorb the moisture from the filling. An additional (but optional) step is to egg wash the pastry all over before returning it to the oven, which acts as a seal and prevents any moisture from entering the crust.

Beginning at the bottom left, follow the arrows up and over for a neat summary of how to correctly blind bake a flaky pie crust for a crisp, buttery, golden-brown pie (with no underbaked, soggy bottom).

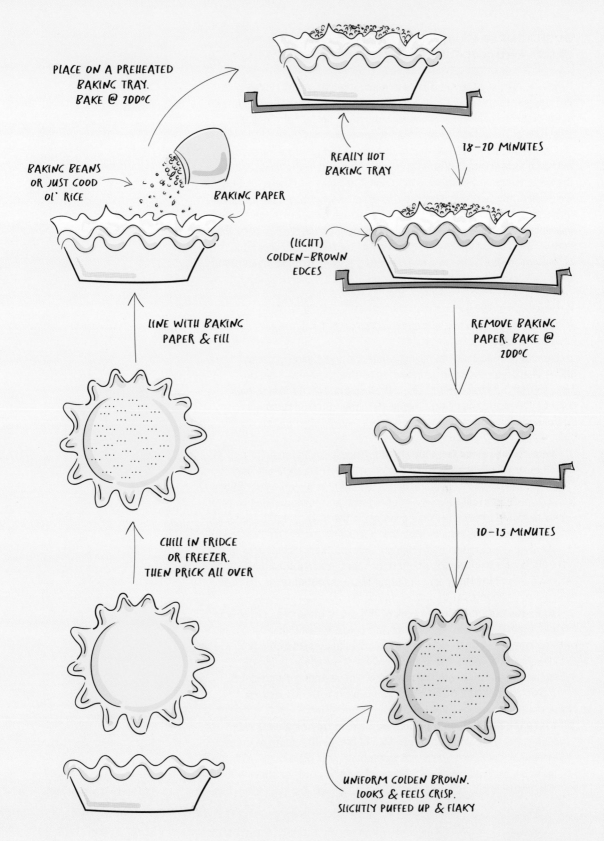

PLACE ON A PREHEATED BAKING TRAY. BAKE @ 200°C

REALLY HOT BAKING TRAY

18-20 MINUTES

BAKING BEANS OR JUST GOOD OL' RICE

BAKING PAPER

(LIGHT) GOLDEN-BROWN EDGES

REMOVE BAKING PAPER. BAKE @ 200°C

LINE WITH BAKING PAPER & FILL

10-15 MINUTES

CHILL IN FRIDGE OR FREEZER. THEN PRICK ALL OVER

UNIFORM GOLDEN BROWN. LOOKS & FEELS CRISP. SLIGHTLY PUFFED UP & FLAKY

OVEN TEMPERATURES FOR FLAKY ALL-BUTTER PIE CRUST + ROUGH PUFF PASTRY

While shortcrust pastry and tarts in general are baked at 180°C, both flaky pie crust and rough puff pastry benefit from a higher temperature of 200°C. With pastry that relies on lamination for its flaky, delicate structure, choosing the correct oven temperature is a thing of balance.

On the one hand, you want a high enough temperature for the water in the butter to evaporate rapidly almost the moment the pastry hits the oven, creating the characteristic rise, flake and 'puffiness' of flaky pie crust or rough puff pastry. On the other hand, you're battling the tendency of gluten-free pastry to dry out, which sets the crumb and prevents any further rise. An oven temperature of 200°C hits that sweet spot – 180°C is too low to give significant rise and flakiness, and 220°C is too high, drying it out and so preventing it from arriving at delicate, flaky deliciousness.

HOW TO PREVENT A SOGGY BOTTOM

If there's one thing that will destroy a pie, it's a sad, floppy, soggy base. That's why I've collected a series of little things you can do to prevent it from happening. The following tips are relevant to pies for which you add the filling directly on to the unbaked pie crust (no blind baking involved).

1. **Reduce the juices from any fruit filling.** It doesn't matter whether it's a blackberry pie (see page 212) or an apple pie (see page 208) or whether you're using fresh or frozen fruit – your filling will inevitably release a whole lot of juice, and the last place you want this to happen is in the oven while the pie is baking. Macerate or pre-cook the filling, draining the juices and reducing them down for about 15 minutes until thick and syrupy, then add them to the fruit along with some cornflour to further thicken and stabilise the filling. The bonus is that all this also concentrates the flavours.

2. **Bake the pie in the lower part of the oven.** Listen, the top of your pie, especially if you egg wash it and because you bake it at 200°C, is guaranteed to bake all the way to crisp, flaky perfection. If you bake your pie on the lower-middle shelf of the oven, you help guarantee success with the base, too. The resulting pie has both a perfectly baked top and a dark, golden-brown, crisp bottom.

3. **Place the pie dish on a preheated baking tray or sheet.** This kick-starts the baking on the bottom of the pie the moment the pie enters the oven, and you're not waiting for your pie dish to heat up sufficiently on all sides. As a bonus, the baking tray or sheet will also catch any stray juices that might bubble up and out of the dish.

The easiest way to prevent a 'soggy bottom' is to place the pie dish on a preheated baking tray or sheet in the oven. The heat from the baking tray penetrates the base of the pie crust so that the crust cooks and crisps up faster than it would from the heat of the oven alone.

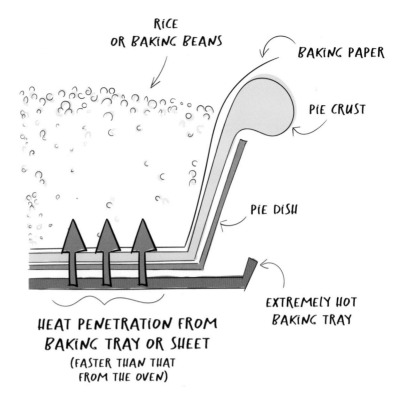

RICE
OR BAKING BEANS

BAKING PAPER

PIE CRUST

PIE DISH

EXTREMELY HOT
BAKING TRAY

HEAT PENETRATION FROM
BAKING TRAY OR SHEET
(FASTER THAN THAT
FROM THE OVEN)

4. **Use a metal pie dish.** I know – controversial. Many bakers prefer glass or ceramic pie dishes, because these supposedly give a more even bake. Glass, of course, has the added advantage that you can actually see what's going on with your pie crust, giving you a better sense of control. But, both glass and ceramic are poor conductors of heat, which pretty much defeats the purpose of doing points #2 and #3 in this list. Glass and ceramic dishes slow down the process of the pie bottom baking, browning and crisping as soon as possible. Metal, on the other hand, heats up in no time and kick-starts the bottom baking pretty much straight away, significantly reducing the likelihood of a soggy crust, even with incredibly juicy fillings (think apple or berry pies). All that said, if you have a glass or ceramic pie dish that you know and love, and it gives consistently excellent results with no soggy bottom in sight, of course go right ahead and use it.

FLAKY ALL-BUTTER PIE CRUST

MAKES 1 quantity
PREP TIME 45 mins
CHILL TIME 1 hour

375g gluten-free flour blend
2 tablespoons caster sugar (for
    sweet pastry) or 1 teaspoon
    caster sugar (for **savoury**)
1½ teaspoons xanthan gum
½ teaspoon salt
250g unsalted butter, cut into
    1cm cubes, chilled
150g cold water ①

NOTES

① *While most wheat-based
pie-crust recipes follow the
3-2-1 rule for the flour:butter:water
ratio, a gluten-free version needs
a 3-2-slightly-more-than-1 rule
owing to the greater water-
absorption capacity of gluten-free
flours. The best ratio for the
gluten-free flour blends in this
book (see page 19) is 3:2:1.2
– but it might be different for
other blend compositions.*

② *Because you leave the butter
pieces relatively large, the dough
is initially (quite unsuccessfully)
kept together only by the added
water, which is why it will look
incredibly dry and crumbly.
As you knead, however, small
portions of the butter melt from
the heat of your hands, helping
the dough to come together.*

The secret to making perfectly flaky, buttery pie crust lies in keeping the butter pieces large and in several rounds of folding to create a rudimentary form of lamination. This, in turn, makes the dough easier to handle and the baked crust extra flaky. When making the pie crust, you may be tempted to add more water. Resist: adding more water will just make the crust tough. Instead, keep kneading and see the pastry transform into a beautiful dough (see Note #2). This recipe makes enough for a 23cm pie with both a base and a full pastry lid.

+ In a large bowl, mix together the gluten-free flour blend, sugar, xanthan gum and salt.
+ Add the cold butter and toss it in the flour until each piece is coated.
+ Squish each cube of butter between your fingers to form small, thin sheets of butter. Be sure to check you've squished all the butter pieces before adding the water.
+ Add the cold water and mix well with a fork or spatula until the dough starts coming together. After you've added all the water, the dough will still look very dry – don't be tempted to add more.
+ Gently squeeze and knead the dough, pressing it against the sides of the bowl, until it comes together in a shaggy ball with little-to-no dry patches (this can take up to 5 minutes). ② If absolutely necessary, and if after 5 minutes the dough still hasn't come together, sprinkle a tiny amount of water on the most persistent dry patches.
+ Shape the dough into a disc, wrap it in cling film and refrigerate it for about 30 minutes or until it feels firm but not hard when you press down on it with your finger.
+ Roll out the chilled dough on a lightly floured surface to a 15 x 45cm rectangle. Turn the dough so that a short end is closest to you. Brush off any excess flour and fold the dough as you would an A4 letter – bottom third up towards the middle and the top third down over it.
+ Turn the dough by 90 degrees (so that the open ends are closest and farthest from you). Roll out into a similar-sized rectangle and repeat the letter fold. For the best results, complete a total of 4–6 letter folds (the more you do, the greater the flakiness). If the dough becomes too soft at any point, chill it in the fridge for 15–30 minutes, then continue.
+ Wrap the dough in cling film and chill it for at least 30 minutes or until needed. (If you chill it for longer than 1 hour before using, soften it for about 15–30 minutes at room temperature, so that it's pliable.)

STORAGE

1 week in the fridge; 2–3 weeks in the freezer. Allow the pastry to thaw at room temperature for at least 1 hour before using – if you press down on it with a finger, you should leave an indentation (but it should feel fairly firm, not too soft to the touch).

ROUGH PUFF PASTRY

MAKES 1 quantity
PREP TIME 30 mins
CHILL TIME 30 mins

225g gluten-free flour blend
½ tablespoon caster sugar
¾ teaspoon xanthan gum
¼ teaspoon salt
40g unsalted butter, softened
100g cold water
125g unsalted butter, frozen
    then coarsely grated

I know that making your own (rough) puff pastry sounds daunting, but I really urge you to give it a try. It's absolutely delicious and actually surprisingly simple and quick to prepare. Instead of using a block of butter that is wrapped in dough, as with traditional puff pastry, rough puff uses grated frozen butter, incorporated through what is essentially a simple letter fold. This makes the butter layer far more malleable and less likely to break during the rolling process, while still creating a neat enough lamination to result in a significant puff and flakiness once baked. A total of six letter folds, as here, creates 486 layers, which is slightly less than the more usual 729 layers of puff pastry, but it's still plenty to give a very impressive end result.

+ In a bowl, combine the gluten-free flour blend, sugar, xanthan gum and salt. Add the softened butter and work it into the dry ingredients until the texture resembles breadcrumbs or coarse sand.
+ Add the water and knead the ingredients to a smooth, firm dough.
+ Roll out the dough on a lightly floured surface to a 15 x 45cm rectangle. Turn the dough so that a short end is closest to you. Brush off any excess flour and place half the grated frozen butter on the middle third of the rectangle, arranging it in an even layer from edge to edge.
+ Start the fold, just as you would an A4 letter: take the bottom third up and over the butter, then gently pat it down.
+ Brush away any excess flour, then place the remaining grated butter on top of the folded dough, again creating an even layer. Fold the top third of dough down over the butter, again gently patting it down.
+ Turn the dough by 90 degrees (so that the open ends are closest and farthest from you). Lightly press down with the rolling pin at several points along the length to ensure a more even distribution of butter.
+ Roll out the dough to a similar rectangle as before. Brush away any excess flour, then fold the bottom third up towards the middle, and the top third down over it – a letter fold. Repeat the sequence of turn, roll and fold 4 more times (a total of 6 folds). Lightly flour the surface and rolling pin as needed. If the pastry becomes too warm (and the butter is too soft), wrap it in cling film, chill it for 15–30 minutes, then continue.
+ After the final fold, wrap the rough puff tightly in cling film and refrigerate it for at least 30–45 minutes or until needed. (If you chill it for longer than 1 hour before using, soften it for about 15–30 minutes at room temperature, so that it's pliable.)

STORAGE

1 week in the fridge; 2–3 weeks in the freezer. Allow the pastry to thaw at room temperature for at least 1 hour before using – if you press down on it with a finger, you should leave an indentation (but it should feel fairly firm, not too soft to the touch).

PLAIN SWEET SHORTCRUST PASTRY (PÂTE SUCRÉE)

MAKES 1 quantity
PREP TIME 15 mins
CHILL TIME 1 hour

225g gluten-free flour blend
¼ teaspoon xanthan gum
¼ teaspoon salt
70g unsalted butter, cubed
  and softened
70g icing sugar
1 egg, at room temperature

I've optimised the process of making this gluten-free shortcrust to give an end result that is crisp and sturdy enough to hold its shape and any fillings, while also being crumbly enough to melt on your tongue. By working softened butter into the flour, you create very tiny pockets of flour-encased butter. This, plus adding the sugar only after you've incorporated the butter, creates the perfect sweet shortcrust pastry – and it does so effortlessly.

+ In a large bowl, using a fork or wooden spoon, combine the gluten-free flour blend, xanthan gum and salt. Mix briefly until all the ingredients are evenly distributed.
+ Add the butter and rub it into the dry ingredients with your fingertips until the texture resembles breadcrumbs or coarse sand.
+ Sift in the icing sugar and mix briefly until evenly distributed.
+ Add the egg and mix until the dough comes together in a ball.
+ Turn out the dough on to the work surface and knead briefly until smooth.
+ Shape the dough into a disc, wrap it in cling film and refrigerate for about 1 hour or until needed. If you refrigerate it for longer than 1 hour, allow the pastry to soften for about 10 minutes at room temperature before using.

STORAGE

3–4 days in the fridge; 1–2 weeks in the freezer. If frozen, allow the pastry to thaw at room temperature for about 30–45 minutes before using – if you press down on it with a finger, you should leave an indentation (but it should feel fairly firm, not too soft to the touch).

CHOCOLATE SWEET SHORTCRUST PASTRY (PÂTE SUCRÉE)

MAKES 1 quantity
PREP TIME 15 mins
CHILL TIME 15 mins

100g dark chocolate (60–65% cocoa solids), chopped ①
70g unsalted butter
220g gluten-free flour blend
10g Dutch processed cocoa powder
½ teaspoon xanthan gum
¼ teaspoon salt
70g icing sugar
1 egg, at room temperature

NOTES

① *I don't recommend using a 70%+ chocolate for this recipe, as it can give a pastry that is too dry and therefore too crumbly to handle.*

② *It's very important that you cool the chocolate–butter mixture all the way to room temperature – a hot or warm mixture will give a crumbly pastry that's difficult to roll out and transfer into a tart tin.*

③ *Because the pastry contains chocolate (and not just cocoa powder), it becomes nearly solid in the fridge as the chocolate sets and crystallises. Even after 1 hour or so at room temperature, it would be too crumbly to roll without kneading. Kneading makes it pliable again (mostly because the heat in your hands warms the chocolate).*

This variation on plain shortcrust earns a space of its own, as it goes beyond just adding cocoa powder. Instead, I incorporate a rather large amount of melted chocolate – and it's pretty much a scientifically proven fact that adding a large amount of chocolate to anything makes it 1,000 times better. Here, the chocolate imparts a wonderfully deep, rich flavour, and ramps up the crispness of the pastry. Because chocolate takes a shorter time to crystallise and solidify in the fridge than butter, the chocolate pastry requires both a shorter refrigeration and a longer 'thawing' time before using than its plain equivalent.

+ Melt the chocolate and butter together in a heatproof bowl over a pan of simmering water. Remove from the heat and allow to cool to room temperature. ②
+ In a separate large bowl, using a fork or wooden spoon, combine the gluten-free flour blend, cocoa powder, xanthan gum and salt. Mix briefly until all the ingredients are evenly distributed.
+ Add the melted chocolate–butter mixture and mix until combined.
+ Sift in the icing sugar and mix briefly until evenly distributed. If the chocolate mixture has firmed up too much to mix easily, rub in the icing sugar using your fingertips. The final texture should resemble breadcrumbs or coarse sand.
+ Add the egg and mix until the dough comes together in a ball.
+ Turn out the dough on to the work surface and knead briefly until smooth.
+ Shape the dough into a disc, wrap it in cling film and refrigerate for about 15–20 minutes or until needed. If you refrigerate it for longer than 1 hour, allow it to soften at room temperature for about 1–2 hours before using. Before rolling, give the pastry a thorough knead until it is perfectly pliable. ③

STORAGE
Maximum 2 days in the fridge. (I don't advise freezing this pastry.)
Allow the pastry to thaw at room temperature for a few hours before using – if you press down on it with a finger, you should leave an indentation. Give it a thorough knead before rolling out.

CARAMEL APPLE PIE

SERVES 8–10
PREP TIME 1 hour 45 mins
(including macerating the apples;
excluding the pastry)
COOK TIME 5 mins
CHILL TIME 30 mins
BAKE TIME 1 hour

1 quantity of **sweet flaky all-
    butter pie crust** (see page 200)
7–8 (about 1.4 kg) sweet–tart, firm
    eating apples (such as Granny
    Smith, Pink Lady or Braeburn)
4 tablespoons lemon juice
200g caster sugar
2 teaspoons ground cinnamon
1 teaspoon ground ginger
¼ teaspoon ground nutmeg
25g cornflour
1 egg, whisked
1–2 tablespoons granulated sugar
vanilla ice cream, to serve
    (optional)

Perfectly crisp, flaky pastry (both on top and on the definitely-not-soggy bottom), a juicy apple filling with delicious caramel notes… welcome to gluten-free, apple-pie heaven. The magic of this recipe comes from reducing the juices released by the apples, which concentrates the flavour and adds wonderful caramel undertones to the filling. At the same time, this removes a lot of the moisture from the filling, drastically reducing the likelihood of a sad and soggy bottom. Macerating the apple slices also softens them slightly, allowing you to pack more fruit into the pie and reducing shrinkage during baking. The baking time is optimised to give a golden-brown, crisp bottom crust and apples that are softened without being mushy, with just the hint of a bite to them.

+ Pre-make the pastry according to the instructions on page 200.
+ Peel, core and quarter the apples, then cut them into approximately 5mm slices and toss with lemon juice (to prevent oxidation and browning). Add the caster sugar and spices, and mix well.
+ Allow to macerate at room temperature for 1–2 hours until the apples are slightly softened and have released their juices. About 30 minutes before you start assembling the pie, transfer the apples into a colander or sieve placed over a bowl to drain and reserve the juice.
+ You'll need a 23cm pie tin (4cm deep). Remove the flaky pie crust from the fridge or freezer to come up to room temperature as directed on page 200 (to prevent cracking).
+ Pour the collected apple juice into a large saucepan and reduce over a medium–high heat, stirring frequently, until syrupy (about 5–10 minutes). Once reduced, set aside to cool until warm.
+ Divide the dough into two unequal portions, in about a 40:60 ratio.
+ On a lightly floured piece of baking paper, roll out the smaller portion of dough to a 23 x 25cm rectangle, about 3mm thick. Cut out 10 strips, each 2.5cm wide and 23cm long, and set aside until needed.
+ On a lightly floured piece of baking paper or cling film, roll out the larger portion of dough to about 3mm thick and cut out a 30cm circle. Transfer the pastry circle to the pie tin, making sure it's snug against the bottom and sides. Trim the excess, leaving about a 2.5cm overhang.
+ Add the cornflour to the apple slices and toss to coat evenly.
+ Transfer the apple slices into the pie crust, making sure the layers are as snug as possible to reduce shrinkage and collapse during baking. Drizzle the slightly cooled, reduced juices over the filling (reheat them briefly if they've thickened too much while cooling).

+ Using the strips of pastry, create a lattice over the filling to form the pie lid. Trim the strips to fit, making sure they sit on top of the pie-crust edge.
+ Fold the overhanging pastry inwards, so that it covers the ends of the lattice strips. Crimp the edges: squeeze the pastry between the thumb and index finger of your non-dominant hand (forming a V-shape) and press gently into the V using the index finger of your dominant hand.
+ Refrigerate the pie for about 30 minutes. ①
+ Line a large baking tray or sheet with a large piece of foil and place it on the lower-middle oven shelf. Preheat the oven to 200°C. ②
+ Once the pie has chilled, glaze it using the egg and sprinkle generously with granulated sugar. Place the pie on the hot baking tray or sheet in the oven and bake for about 1 hour–1 hour 15 minutes, until the juices bubble up through the lattice and the internal temperature of the filling reaches about 90°C. If the pie starts browning too quickly, cover it with foil (shiny side up) and bake until done (uncover again for the last 5–7 minutes for an extra-crisp pie lid).
+ Allow to cool for at least 4 hours before serving, with a scoop of vanilla ice cream, if you wish. ③

STORAGE

Best on the day, but you can keep the pie lightly covered with kitchen paper for up to 2 days. Don't use a closed container, as that would significantly soften the crust.

NOTES

① *Chilling the pie sets the butter in the crust, which helps the crust keep its shape during baking, prevents butter leakage and gives a perfectly flaky pie.*

② *In addition to preventing a soggy bottom (see pages 198–9), the hot baking tray or sheet catches any juices that bubble up and out of the pie during baking.*

③ *While the pie is delicious warm, the filling requires a few hours to set completely. Before that, the filling will be loose and runny, and cutting into the pie will be a messy affair.*

PHOTOGRAPHY OVERLEAF ⟶

BLACKBERRY PIE

SERVES 8–10
PREP TIME 45 mins
(excluding the pastry)
COOK TIME 15 mins
CHILL TIME 30 mins
BAKE TIME 50 mins

1 quantity of **sweet flaky all-butter pie crust** (see page 200)
900g fresh or frozen blackberries ①
300–350g caster sugar, depending on the sweetness of the blackberries
4 tablespoons lemon juice
25g cornflour
1 egg, whisked
1–2 tablespoons granulated sugar
vanilla ice cream, to serve (optional)

The secret to making a perfectly juicy, aromatic blackberry pie with a crisp, flaky pie crust lies in pre-cooking the filling – regardless of whether you're using fresh or frozen fruit. Pre-cooking involves cooking the blackberries until they're softened and have released their juices, and then draining and reducing those juices further until syrupy (when hot) or jam-like (when cooled). This process minimises the likelihood of a soggy crust, concentrates the delicious berry tang and eliminates the need for large quantities of cornflour, which can dilute the flavour and give the filling an unpleasant floury or gummy texture. You can use this method for other berries, as well as stone fruit such as cherries or apricots.

+ Pre-make the pastry according to the instructions on page 200.
+ You will need a 23cm pie tin (4cm deep). Remove the flaky pie crust from the fridge or freezer to come up to room temperature as directed on page 200 (to prevent cracking).
+ In a large saucepan, combine the blackberries, caster sugar and lemon juice, and cook over a medium–high heat until the blackberries have softened and released their juices. Drain the blackberries through a sieve set over a bowl to extract the juice, and return the juice to the pan. Tip the blackberries into a bowl and set aside.
+ Reduce the juice over a medium–high heat, stirring frequently, until they have thickened and become syrupy (about 15 minutes). Once syrupy, remove from the heat and allow to cool until warm.
+ Add the cornflour to the whole blackberries and mix to coat evenly. Then, add the cooled, reduced juice and mix to combine. Set aside until needed.
+ Divide the dough into two unequal portions, in about a 40:60 ratio.
+ On a lightly floured piece of baking paper, roll out the smaller portion of dough to a 23 x 25cm rectangle, about 3mm thick. Cut out 10 strips, each 2.5cm wide and 23cm long, and set aside until needed.
+ On a lightly floured piece of baking paper or cling film, roll out the larger portion of dough to about 3mm thick and cut out a 30cm circle. Transfer the pastry circle to the pie tin, making sure it's snug against the bottom and sides. Trim the excess, leaving about a 2.5cm overhang.
+ Pour in the blackberry filling, smoothing it out into an even layer.
+ Using the strips of pastry, create a lattice over the filling to form the pie lid. Trim the strips to fit, making sure they sit on top of the pie-crust edge.

+ Fold the overhanging pastry inwards to cover the ends of the lattice strips. Crimp the edges: squeeze the pastry between the thumb and index finger of your non-dominant hand (forming a V-shape) and press gently into the V using the index finger of your dominant hand.
+ Refrigerate the pie for about 30 minutes. ②
+ Line a large baking tray or sheet with a large piece of foil and place it on the lower-middle oven shelf. Preheat the oven to 200°C. ③
+ Once the pie has chilled, glaze it using the egg and sprinkle generously with granulated sugar. Place the pie on the hot baking tray or sheet in the oven and bake for about 50–60 minutes until the juices bubble up through the lattice and the internal temperature of the filling reaches about 90°C. If the pie starts browning too quickly, cover it with foil (shiny side up) and bake until done (uncover again for the last 5–7 minutes for an extra-crisp pie lid).
+ Allow to cool for at least 3–4 hours before serving, with a scoop of vanilla ice cream, if you wish. ④

STORAGE

Best on the day, but you can keep the pie lightly covered with kitchen paper for up to 2 days. Don't use a closed container as that would significantly soften the crust.

NOTES

① *This seems like a lot, but because the filling is pre-cooked, you can pack it more tightly resulting in a beautifully tall pie with minimal shrinkage during baking.*

② *Chilling the pie sets the butter in the crust, which helps the crust keep its shape during baking, prevents butter leakage and gives a perfectly flaky pie.*

③ *In addition to preventing a soggy bottom (see pages 198–9), the hot baking tray or sheet catches any juices that bubble up and out of the pie during baking.*

④ *The long cooling time is necessary for the filling to set properly. Cutting the pie too soon can result in a juicy and runny (if delicious) mess.*

PHOTOGRAPHY OVERLEAF ⟶

LEMON MERINGUE PIE

SERVES 8–10
PREP TIME 45 mins
(excluding the pastry)
CHILL TIME 4 hours 15 mins
BAKE TIME 28 mins
COOK TIME 6 mins

½ quantity of **sweet flaky
all-butter pie crust**
(see page 200)

LEMON FILLING

190g water
75g lemon juice
150–200g caster sugar
25g cornflour
¼ teaspoon salt
4 egg yolks
40g unsalted butter

SWISS MERINGUE

4 egg whites
150g caster sugar
¼ teaspoon cream of tartar

This isn't a traditional way to make lemon meringue pie – the traditional way is strewn with steps seemingly designed to trip you up, so that you're left with a runny, gloopy filling or a weeping meringue (and consequently a weeping you, at the sad state of a pie that looked oh-so promising). In this version, the only thing baked is the pie crust. The filling is fully cooked on the stovetop, ensuring that it's stable and cuts into beautifully neat slices. The topping is a Swiss meringue, easy to whip up and toast to perfection with a kitchen blow torch just before serving. The end result has all the flavour and the looks of a classic lemon meringue pie... just with far fewer hazards.

PIE CRUST

+ Pre-make the pastry according to the instructions on page 200.
+ You'll need a 23cm pie tin (3cm deep). Remove the flaky pie crust from the fridge or freezer to come up to room temperature as directed on page 200 (to prevent cracking).
+ On a lightly floured piece of baking paper or cling film, roll out the dough to 3mm thick and cut out a rough 30cm circle. Transfer the pastry circle to the pie dish, making sure it's snug against the bottom and sides. Trim the excess dough, leaving about a 2.5cm overhang.
+ Fold the overhanging dough under itself. Crimp the edges: squeeze the pastry between the thumb and index finger of your non-dominant hand (forming a V-shape) and press gently into the V using the index finger of your dominant hand.
+ Refrigerate the pastry lining for 15–30 minutes. ①
+ Place a large baking tray or sheet on the lower-middle oven shelf and preheat the oven to 200°C.
+ Once the pie crust has chilled, prick the base all over with a fork, line it with a scrunched-up piece of baking paper and fill with baking beans or rice.
+ Place the pie on the hot baking tray or sheet and bake for 18–20 minutes until you can see the edges turning a light golden colour.
+ Remove the crust from the oven and take out the baking beans or rice and baking paper. Return the uncovered pie crust to the oven for a further 10–15 minutes until crisp and golden brown.
+ Place the baked pie crust in the tin on a wire rack to allow the crust to cool completely.

LEMON FILLING

+ In a saucepan, whisk together the water, lemon juice, sugar, cornflour and salt. Add the egg yolks and whisk lightly until combined.
+ Cook over a medium–high heat, stirring continuously, for 4–5 minutes until thickened and the mixture starts bubbling. Then, cook for a further 2 minutes, stirring vigorously throughout. ②
+ Remove from the heat and stir in the butter until melted.
+ Pour the filling into the cooled pie crust. Cool at room temperature, then refrigerate for at least 4 hours or preferably overnight.

SWISS MERINGUE

+ Prepare the Swiss meringue topping as close to serving as possible. Mix the egg whites, sugar and cream of tartar in a heatproof bowl over a pan of simmering water. Stir continuously until the meringue mixture reaches 65°C and the sugar has dissolved.
+ Remove the mixture from the heat and transfer it to a stand mixer with the whisk attachment, or use a hand mixer fitted with the double beaters, and whisk on medium–high speed for 5–7 minutes until greatly increased in volume and stiff peaks form.
+ Top the chilled lemon pie with the Swiss meringue. Lightly toast the meringue using a kitchen blow torch, then slice and serve.

STORAGE

Refrigerate the filled pie without the Swiss meringue, and covered with cling film, for up to 2 days. Make and toast the meringue topping just before serving.

NOTES

① *Chilling the pie crust sets the butter, which helps the crust keep its shape during baking, prevents butter leakage and gives a perfectly flaky pie. It also prevents any damage to the crust as you line it with the baking paper for blind baking.*

② *In order to harness the entire thickening power of cornflour, it's important to bring the filling to the boil and keep it there for a minute or two. That's because 'starch gelatinisation' (the process during which the bonds between starch molecules break down in favour of starch–water bonds, resulting in gel formation) requires temperatures of 100°C (boiling point of water) or above.*

PHOTOGRAPHY OVERLEAF ⟶

TARTE TATIN

SERVES 6–8
PREP TIME 45 mins
(excluding the pastry)
COOK TIME 5 mins
BAKE TIME 1 hour

½ quantity of **sweet flaky
all-butter pie crust**
(see page 200)
150g caster sugar
50g unsalted butter
5–6 sweet–tart, firm eating apples
(such as Granny Smith, Pink
Lady or Braeburn), peeled, cored
and quartered
cream or vanilla ice cream, to serve
(optional)

NOTES

① *Although the apple juices
and caramel are fairly unlikely
to overflow while baking,
I recommend you place the pie
tin on to a large baking tray or
sheet to catch any stray liquids
and prevent them from burning
on the bottom of the oven (and
smoking out your kitchen).*

② *I don't recommend you wait for
more than 10–15 minutes before
turning out the tarte tatin on to
a plate, as the caramel will start
to cool and set, resulting in the
apples sticking to the pie tin.*

I've tested numerous ways of preparing tarte tatin (mostly to do
with what you do to the apples) and this one is by far my favourite.
Not only is it about as fuss-free as it gets, it also produces the most
consistently mouthwatering results. Baking the apples uncovered for
25 minutes ensures that some of the water they release evaporates
before the pastry goes on and the remaining juices concentrate into
sticky deliciousness. The results: no soggy pastry or apple mushiness
– just crisp, flaky pastry, tender apples and caramel goodness.

+ Pre-make the pastry according to the instructions on page 200.
+ You'll need a 23cm pie tin. Adjust the oven shelf to the middle position,
  preheat the oven to 180°C and remove the flaky pie crust from the
  fridge or freezer to come up to room temperature as directed on page
  200 (to prevent cracking).
+ In a saucepan, cook the sugar over a medium heat for 5–8 minutes
  until completely melted and a deep amber colour.
+ Remove from the heat, add the butter and stir well until fully melted.
  Pour the caramel into the pie tin.
+ Arrange the apple quarters face downwards on top of the caramel layer,
  fanning them in concentric circles, beginning around the outside edge
  of the pie tin and working inwards to cover the caramel. Make sure to
  pack the apples snugly together, almost overcrowding them, as they
  will shrink during baking.
+ Place the pie tin on a large baking tray or sheet and bake the apples,
  without a pastry lid for 25 minutes. ① At this point, they will have
  released some of their juices and will be slightly softened.
+ While the apples are baking, roll out the pastry to about 4mm thick
  and cut out a circle slightly larger than your pie dish. Return the pastry
  circle to the fridge until needed.
+ After 25 minutes in the oven, remove the pie tin and cool the filling at
  room temperature for about 10 minutes – this will ensure that the pastry
  doesn't melt on contact with the hot apples and caramel.
+ Then, cover the apples with the pastry circle, tuck in the edges, make a
  small hole or slit in the middle to help release steam during baking, and
  return the dish to the oven. Bake the tatin for 35–40 minutes or until the
  pastry is puffed up and golden brown on top.
+ Remove from the oven and cool for 10–15 minutes. Then, place a plate
  slightly larger than the pie dish on top and invert the tart on to it. ②
+ Serve warm, with cream or a scoop of ice cream, if you wish.

STORAGE

Best on the day, but you can keep it lightly covered with kitchen paper for
up to 2 days. Don't use a closed container as that would soften the crust.

APPLE + MARZIPAN GALETTE

SERVES 6–8
PREP TIME 45 mins
(excluding the pastry)
CHILL TIME 15 mins
BAKE TIME 45 mins

½ quantity of **sweet flaky
all-butter pie crust**
(see page 200)

MARZIPAN FRANGIPANE
150g marzipan
100g almond flour
70g unsalted butter, softened
3 tablespoons caster sugar
2 egg yolks, at room temperature

TO ASSEMBLE
3–4 firm eating apples with a red
or pink skin (such as Pink Lady),
cored and quartered ①
juice of ½ lemon
1 egg, whisked
20g flaked almonds
2 tablespoons granulated sugar

NOTE
① *Although you could use green
apples such as Granny Smith for
this recipe, the galette will look
much prettier with reddish-pink
apples with occasional patches
of yellow or orange.*

A galette has all the benefits of a pie – the flaky, buttery, golden-brown pie crust and a delicious filling – with none of the fuss. There's no pie tin or lattice-making or crimping here, just roll out the pastry, cut out a rough circle, pile on the filling, fold in the pastry edges, give them a quick egg wash, and into the oven it goes. This galette features an abundance of thinly sliced, tart apples arranged on top of the most delicious, aromatic frangipane made from marzipan. It is a flavour match made in heaven, if there ever was one.

+ Pre-make the pastry according to the instructions on page 200.
+ Remove the flaky pie crust from the fridge or freezer to come up to room temperature as instructed on page 200 (to prevent cracking).
+ Combine all the marzipan frangipane ingredients in the bowl of a food processor or blender, and blitz until you get a smooth paste. Set aside until needed.
+ Cut the apple quarters (skin on) into 3mm slices, keeping the slices grouped together in their quarters. Pour over the lemon juice (to prevent oxidation and browning) and set aside until needed.
+ On a lightly floured piece of baking paper, roll out the flaky pie crust to 3mm thick and cut out a 30cm circle. Spread the frangipane in an even layer on top, leaving a 5cm border around the edge.
+ Arrange the sliced apple quarters on top of the frangipane, crowding them closely together.
+ Fold the pastry border over the apples and press down gently to seal.
+ Trim the excess baking paper and transfer the galette, together with the baking paper, on to a large baking sheet. Refrigerate for at least 15 minutes.
+ Adjust the oven shelf to the middle position and preheat the oven to 200°C.
+ Brush the chilled pastry with the whisked egg, sprinkle the galette with the flaked almonds and granulated sugar, and bake for 45–50 minutes or until the pastry is deep golden brown and crisp. Using a large offset spatula, very carefully lift the pastry to check the bottom – it should be golden brown too. If the pastry starts browning too quickly, cover the galette with foil (shiny side up) and bake until done.
+ Cool on the baking sheet for 10–15 minutes, then transfer to a wire rack to cool completely.

STORAGE
Best on the day, but you can keep it lightly covered with kitchen paper for up to 2 days. Don't use a closed container as that would soften the crust.

TRIPLE CHOCOLATE SWEET PASTRIES

MAKES 10
PREP TIME 1 hour
(excluding the pastry)
COOK TIME 5 mins
BAKE TIME 2 x 22 mins

Chocolate pastry, chocolate fudge filling, chocolate icing and chocolate sprinkles on top – talk about a chocolate overload! It's pretty impossible not to love these sweet pastries, and they're incredibly fun to make. Plus, when you take that first bite through the crisp, crumbly crust and the luscious chocolate fudge filling hits your taste buds... well, it's love at first bite.

SWEET PASTRIES

1 quantity of **sweet flaky all-butter pie crust** (see page 200), with 20g gluten-free flour blend substituted with 40g Dutch processed cocoa powder ①
1 egg, whisked
chocolate sprinkles, to decorate

CHOCOLATE FUDGE FILLING

75g whole milk
65g caster sugar
35g Dutch processed cocoa powder
¼ teaspoon salt
125g dark chocolate (60–70% cocoa solids), chopped
50g unsalted butter

CHOCOLATE ICING

100g icing sugar
25g Dutch processed cocoa powder
3–4 tablespoons water

NOTE

① *While cocoa powder is a dry ingredient, it contains a significant amount of fat (unlike the gluten-free flour blend, which mainly comprises starch and proteins). Consequently, I don't recommend a 1:1 substitution with flour.*

CHOCOLATE FUDGE FILLING

+ In a saucepan over a medium–high heat, combine the milk, caster sugar, cocoa powder, salt and half the chocolate, until melted.
+ Remove from the heat and add the butter and the remaining chocolate. Mix until the butter and chocolate have melted and you get a smooth, glossy mixture. (This should take about 5 minutes altogether.)
+ Set aside until needed. (The filling will firm up slightly on cooling.)

ASSEMBLING THE SWEET PASTRIES

+ Pre-make the pastry according to the instructions on page 200.
+ Adjust the oven shelf to the middle position, preheat the oven to 200°C, line two baking sheets with baking paper, and remove the pastry from the fridge or freezer to come up to room temperature as directed on page 200 (to prevent cracking).
+ Roll out the pastry to about 3mm thick and cut out a 38 x 40cm rectangle. Divide the pastry into 20 smaller rectangles of 7.5 x 10cm.
+ Use the whisked egg to wash the edges of half the pastry rectangles, and spoon a heaped tablespoon of filling into the centre of each one. Spread the filling slightly, but leave a 2cm border around the edge.
+ Place the remaining pastry rectangles over the filling, seal the edges with your fingers and crimp with a fork.
+ Egg wash the tops of the pastries, make a few slits with a knife to allow any steam to escape, and transfer the pastries to the baking sheets.
+ One baking sheet at a time, bake the pastries for 22–24 minutes until puffed up and the pastry feels crisp when you tap it. Cool on a wire rack.

CHOCOLATE ICING

+ Sift together the icing sugar and cocoa powder into a bowl to combine. Add the water, ½ tablespoon at a time, until you get a slightly runny icing consistency that thickly coats the back of a spoon.
+ Spread the icing on top of the cooled sweet pastries and decorate with chocolate sprinkles. Allow to set at room temperature for 30 minutes before serving.

STORAGE

3–4 days in a closed container at room temperature.

CARAMELISED ONION + BROCCOLI QUICHE

SERVES 4–6
PREP TIME 45 mins
(excluding the pastry)
COOK TIME 13 mins
CHILL TIME 15 mins
BAKE TIME 1 hour

½ quantity of **savoury flaky
  all-butter pie crust**
  (see page 200)
3 tablespoons olive oil
3 red onions, thinly sliced
1 medium or 2 small broccoli
  heads, cut into bite-sized florets
  (about 250g prepared weight)
100g cheddar cheese, grated
6 eggs, at room temperature
120g whole milk, at room
  temperature
salt and pepper, to season

NOTES

① *Pressing the pastry into the
  grooves of the tart tin is a
  small detail, but that extra
  few minutes' work gives a
  much more polished finish
  to the baked quiche.*

② *The filling can sometimes puff
  up significantly during baking
  because of the large number of
  eggs, but it always levels out into
  a beautifully flat quiche after
  cooling for about 5–10 minutes.*

There are many delicious components to this vegetarian quiche, but
what really makes it sing are the caramelised onions. Their sweet
flavour melds beautifully with the cheesy filling. The pie crust is
perfectly flaky, and the broccoli is tender, with the slightest bite to it.

+ Pre-make the pastry according to the instructions on page 200.
+ You'll need a 23cm loose-bottomed, fluted tart tin (about 3.5cm deep).
  Remove the pie crust from the fridge or freezer to come up to room
  temperature as directed on page 200 (to prevent cracking).
+ Heat 2 tablespoons of the oil in a saucepan over a medium heat. Add
  the onions, season with salt and pepper and cook for 10–15 minutes
  until soft and caramelised. Transfer to a plate and cool until needed.
+ Add the remaining oil to the hot pan, then add the broccoli florets and
  about 2 tablespoons of water. Season with salt and pepper and cover
  with a lid. Cook for 3–5 minutes, stirring occasionally, until the broccoli
  is tender but not soft. Set aside to cool until needed.
+ On a lightly floured piece of baking paper or cling film, roll out the
  dough until 3mm thick and cut out a roughly 28cm circle. Transfer the
  pastry circle into the tart tin, making sure it's snug against the bottom
  and sides. Roll a rolling pin across the tart tin to trim away any excess,
  and gently press the pastry into the grooves of the fluted edge. ①
  Refrigerate for at least 15 minutes.
+ Place a large baking tray or sheet on the lower-middle oven shelf and
  preheat the oven to 200°C.
+ Once the pie crust has chilled, prick the base with a fork, line it with a
  scrunched-up piece of baking paper and fill with baking beans or rice.
+ Place the pie crust on the hot baking tray or sheet and bake for
  18–20 minutes until you can see the edges turning a light golden colour.
+ Remove the baking beans or rice and the baking paper from the crust
  and return to the oven for a further 7–10 minutes until light golden
  brown and dry to the touch.
+ Remove from the oven, then reduce the oven temperature to 180°C.
+ Mix the onions and broccoli together and spoon them into the crust
  in an even layer. Scatter the cheddar on top. Whisk together the eggs
  and milk to combine and pour them evenly over the filling.
+ Bake the quiche for 35–38 minutes until the filling has puffed up
  slightly, isn't wobbly when shaken and is golden brown on top. ②
+ Cool in the tart tin for 10–15 minutes, then remove from the tin, slice
  into wedges and serve warm.

STORAGE
Best straight after baking, or on the day. Or, keep in a closed container until
the following day and reheat in the microwave for about 30 seconds.

ROASTED BUTTERNUT SQUASH + CHEDDAR FLAKY PASTRIES

MAKES 12
PREP TIME 45 mins
(excluding the pastry)
BAKE TIME 45 mins +
2 x 30 mins
COOK TIME 10 mins
CHILL TIME 15 mins

1 quantity of **savoury flaky
all-butter pie crust**
(see page 200)
1 small butternut squash, peeled,
deseeded and cut into about
1cm chunks (about 500g,
prepared weight)
3 tablespoons olive oil
2 red onions, thinly sliced
50–75g cheddar cheese, grated
2–3 thyme sprigs, leaves picked
pinch of ground nutmeg
1 egg, whisked
mixture of sesame, sunflower and
pumpkin seeds, for sprinkling
salt and pepper, to season

Open a dictionary, look up 'comfort food' and you should (if all were right in the world) find a picture of these butternut squash pastries. The flaky and crisp pastry is the perfect contrast to the melt-in-the-mouth filling of roasted squash, caramelised onions and cheddar. Every bite is an explosion of flavour, and it's pretty much impossible to not come back for a second helping (and possibly a third).

+   Pre-make the pastry according to the instructions on page 200.
+   Adjust the oven shelf to the middle position, preheat the oven to 200°C and line two large baking sheets with baking paper.
+   In a large bowl, toss the butternut squash with 1 tablespoon of the oil and a pinch each of salt and pepper. Transfer the chunks to a lined baking sheet and spread them out in a single layer. Roast for 45–60 minutes until soft and caramelised. Set aside to cool until needed.
+   Remove the flaky pie crust from the fridge or freezer to come up to room temperature as directed on page 200 (to prevent cracking).
+   Heat the remaining olive oil in a saucepan over a medium heat. Add the onions, season with salt and pepper, and cook until caramelised (about 10–15 minutes). Remove from the heat and allow to cool slightly.
+   In a bowl, lightly mash the roasted butternut squash until you get a mixture of chunks and mash. Mix in the caramelised onions, along with the cheddar, thyme and nutmeg. Season with salt and pepper.
+   Roll out the pastry until 2–3mm thick and cut out a 30 x 60cm large rectangle, then cut the pastry into 24 rectangles of 7.5 x 10cm.
+   Egg wash the edges of half of the rectangles and spoon about 1½–2 tablespoons of the filling into the centre of each one. Spread out the filling slightly, leaving a 2cm border around the edge.
+   Place the remaining pastry rectangles on top of the filling, seal the edges with your fingers and crimp with a fork. Transfer the pastries to two lined baking sheets (re-use the one you used for the squash, with a fresh piece of baking paper) and refrigerate for 15–20 minutes.
+   While the pastries are chilling, adjust the oven shelf to the middle position and preheat the oven to 200°C.
+   Egg wash the tops of the pastries and make a slit in each to allow any steam to escape. Sprinkle with the seed mixture. One baking sheet at a time, bake the pastries for about 30–35 minutes or until puffed up, crisp and golden. Cool slightly on a wire rack, then serve warm.

STORAGE
Best immediately after baking, or on the day. Or, keep them in a closed container until the following day and reheat in the microwave for about 30 seconds before serving.

APRICOT 'DANISH' PASTRIES

MAKES 8
PREP TIME 45 mins
(excluding the pastry)
COOK TIME 5 mins
CHILL TIME 10 mins
BAKE TIME 24 mins

Traditionally, these apricot pastries are made with a laminated brioche dough (so-called Danish pastry) – but this version made with rough puff pastry is just as delicious. There's something special about the combination of vanilla-scented pastry cream, apricot halves and buttery, delicate pastry that takes me straight back to my childhood. The fact that they make my kitchen smell like a bakery is a definite bonus.

1 quantity of **rough puff pastry**
(see page 203)

PASTRY CREAM
250g whole milk
1 teaspoon vanilla bean paste
3 egg yolks
75g caster sugar
25g cornflour
25g unsalted butter

PASTRIES
16 apricot halves, canned in syrup
1 egg, whisked
2 tablespoons apricot jam, heated
    until runny (optional), to finish
icing sugar, for dusting

NOTES
① *Whisking the egg yolks and sugar together until smooth is called 'blanching'. The sugar protects the egg proteins, preventing lump formation during cooking.*

② *This is called tempering, and it prevents the egg yolks from scrambling, which would happen if you added the mixture all at once to the hot milk. Tempering slowly increases the temperature of the egg yolks while at the same time diluting them, to give a silky-smooth pastry cream.*

+ Pre-make the pastry according to the instructions on page 203.
+ Adjust the oven shelf to the middle position, preheat the oven to 200°C, line a baking sheet with baking paper, and remove the rough puff pastry from the fridge or freezer to come up to room temperature as directed on page 203 (to prevent cracking).
+ In a saucepan, cook the milk and vanilla paste over a medium–high heat until boiling. In the meantime, whisk together the egg yolks and sugar until pale. ① Add the cornflour and whisk until combined.
+ Pour the hot milk in a slow stream into the egg yolks, whisking continuously. ② Return the mixture to the pan and cook over a high heat, whisking continuously again, until thickened (about 1 minute).
+ Remove from the heat, stir in the butter until melted, then allow to cool completely, stirring or whisking occasionally to prevent a skin forming.
+ Roll out the pastry to about 3mm thick and cut out a 20 x 40cm rectangle. Divide the pastry into 8 squares of 10cm.
+ Spoon about 2 tablespoons of the cooled pastry cream in the middle of each square and place 2 apricot halves on top (cut sides downwards) – one in the upper left corner and one in the lower right corner.
+ Fold the lower left corner towards the middle, partially covering the apricot halves. Egg wash the upper right corner and fold it towards the middle, so that it overlaps the folded pastry, pressing down lightly to seal. Be careful not to fold too tightly, which may squeeze out some of the pastry cream. You shouldn't be able to see any pastry cream, only the apricots. Repeat with the other 7 squares, and transfer them to the lined baking sheet. Refrigerate for 10–15 minutes.
+ Once the pastries have chilled, egg wash the top of the pastry and the visible edges around the apricot halves.
+ Bake for 24–26 minutes until puffed up and an even golden brown, then remove from the oven and transfer to a wire rack to cool.
+ Once the pastries are cold, brush the tops with the runny apricot jam, if you wish, and dust with icing sugar.

STORAGE
Best on the day, but you can keep them in a closed container for up to 2 days.

STRAWBERRY LEMONADE TARTLETS

MAKES 6
PREP TIME 45 mins
(excluding the pastry)
COOK TIME 5 mins
CHILL TIME 10 mins
BAKE TIME 20 mins

1 quantity of **rough puff pastry**
(see page 203)

LEMON PASTRY CREAM
325g whole milk
½ teaspoon vanilla bean paste
4 egg yolks
100g caster sugar
35g cornflour
35g unsalted butter
zest of 1 unwaxed lemon
4 tablespoons lemon juice

TO ASSEMBLE
1 egg, whisked
12 strawberries (6 sliced, 6 halved)
1 tablespoon lemon juice
3–4 tablespoons caster sugar
25g flaked almonds, toasted

NOTE
① *A relatively high oven
temperature of 200°C triggers
the rapid evaporation and
consequent expansion of water
vapour from the butter layers,
resulting in a dramatic rise
and a tender pastry that crackles
most enticingly when you bite
into it. (You can read why
increasing the temperature
to 220°C is a rather bad idea
on page 198.)*

I simply had to include these adorable little tartlets – not only because they're terribly delicious, but because the recipe perfectly shows off the gluten-free rough puff in all its flaky glory. Add to this the lemon-zest-scented pastry cream and macerated strawberries, and you've got a dessert that will set off fireworks on your taste buds.

LEMON PASTRY CREAM
+ In a saucepan, heat the milk and vanilla paste together until boiling. In the meantime, whisk together the egg yolks, sugar and cornflour.
+ Pour the hot milk in a slow stream into the egg yolks, whisking continuously. Return the mixture to the saucepan and cook over a high heat, whisking continuously, until thickened (about 1–2 minutes).
+ Remove from the heat and stir in the butter and lemon zest until the butter has melted. Cool, stirring occasionally to prevent a skin forming.
+ Once cooled to room temperature, whisk in the lemon juice. Set aside.

PASTRY BASE
+ Pre-make the pastry according to the instructions on page 203.
+ Adjust the oven shelf to the middle position and preheat the oven to 200°C. ① Line a baking sheet with baking paper and remove the rough puff pastry from the fridge or freezer to come up to room temperature as directed on page 203 (to prevent cracking).
+ Roll out the pastry to about 3mm thick and cut out a 30cm square. Divide into 6 rectangles of 10 x 15cm and transfer them to the baking sheet. Score each rectangle with a sharp knife, about 1cm in from the outer edge, marking out an 8 x 13cm inner rectangle. Take care not to cut all the way through. Prick the inner rectangles all over with a fork to prevent them rising too much in the oven. Chill for 10–15 minutes.
+ Once the pastry has chilled, brush a 1cm border with the whisked egg.
+ Bake for about 20–24 minutes, until puffed up and the edges are an even golden brown. Transfer to a wire rack to cool.
+ If the inner rectangles have significantly risen during baking, cut along the previously scored rectangle with a sharp knife and press the centres down gently to create space for the pastry cream filling.

ASSEMBLING THE TARTLETS
+ Assemble the tartlets a maximum of 1 hour before serving, otherwise the lemon cream can make the pastry soggy.
+ Place all the strawberry pieces in a bowl and add the lemon juice and sugar. Mix well and allow to macerate for about 10 minutes.
+ Whip the cooled pastry cream until smooth, then divide it between the tartlets, swirling it in an even layer in the inner rectangles. Top with the strawberries, drizzling over the juices in the bowl. Sprinkle on the flaked almonds and serve as soon as possible, or within 1 hour.

CARAMELISED ONION + CHERRY TOMATO TARTLETS

MAKES 4
PREP TIME 45 mins
(excluding the pastry)
COOK TIME 10 mins
CHILL TIME 15 mins
BAKE TIME 40 mins

1 quantity of **rough puff pastry**
  (see page 203)
4 tablespoons olive oil
3 red onions, thinly sliced
1–2 tablespoons roughly chopped
  rosemary, oregano and basil
  leaves
½ tablespoon balsamic vinegar
36 cherry tomatoes
75g cheddar cheese, grated
1 egg, whisked
sesame seeds, for sprinkling
salt and pepper, to season

NOTES

① *The secret to perfectly caramelised onions is nothing more than time, patience, stirring, a watchful eye – and a generous pinch of salt. Adding salt at the start slows down the browning, as it draws out moisture, but results in a richer flavour and even caramelisation. Remember: caramelised onions aren't simply brown onions – they are soft, sweet onions that melt on your palate.*

② *Chilling sets the butter in the rough puff pastry, ensuring an impressive puff in the oven and a deliciously flaky pastry.*

These savoury little tarts will take your breath away. From the juicy cherry tomatoes nestled on a bed of grated cheddar and caramelised onions, to the buttery, flaky, golden-brown rough puff pastry, every mouthful is an explosion of flavour and texture.

+ Pre-make the pastry according to the instructions on page 203.
+ Line a baking sheet with baking paper and remove the rough puff pastry from the fridge or freezer to come up to room temperature as directed on page 203 (to prevent cracking).
+ Heat 2 tablespoons of the olive oil in a frying pan over a medium heat. Add the onions and season with salt and pepper. Cook for 10–15 minutes, stirring frequently, until soft and caramelised. ①
+ Once they have caramelised, remove the onions from the heat and mix in the herbs. Allow to cool slightly.
+ In a large bowl, combine the remaining olive oil with the balsamic vinegar and season with salt and pepper. Add the cherry tomatoes and toss until completely coated.
+ Roll out the pastry to about 3mm thick and cut out a 32cm square. Cut the pastry into 4 squares of 16cm and transfer them to the lined baking sheet.
+ Top each square with equal amounts of the caramelised onions, leaving about a 2cm border around the edge. Top with grated cheddar and 9 coated cherry tomatoes per tart (reserve the balsamic and oil left in the bowl). Fold the edges of the pastry so they slightly cover the outermost tomatoes. Refrigerate for 15–20 minutes. ②
+ While the tarts are chilling, adjust the oven shelf to the middle position and preheat the oven to 200°C.
+ Use the whisked egg to wash the pastry edges, then sprinkle the tarts with sesame seeds. Just before baking, drizzle the reserved balsamic oil over the filling.
+ Bake the tarts for about 40–45 minutes or until the pastry edges are puffed up and golden brown, and the tomatoes have softened and burst.
+ Transfer to a wire rack to cool slightly, then serve.

CHEESE TWISTS

MAKES 25
PREP TIME 45 mins
(excluding the pastry)
CHILL TIME 15 mins
BAKE TIME 2 x 15 mins

Flaky, crispy, cheesy and buttery – these cheese twists are a great snack or dinner-party appetiser. By twisting two pastry strands around each other (rather than just twisting a single strand around itself), you're effectively creating braids. And while braiding takes only a smidgen of additional effort, the end results look neater and more elegant, like golden plaits of cheesy deliciousness.

1 quantity of **rough puff pastry**
  (see page 203)
1 egg, whisked
75–100g parmesan cheese, grated
pinch of cayenne pepper (optional)

NOTES

① *In addition to looking pretty, twisting 2 strands together results in straighter cheese twists that don't bend during baking (which a single twisted strand might do).*

② *Chilling sets the butter in the rough puff pastry, ensuring an impressive puff in the oven and a deliciously flaky pastry.*

+ Pre-make the pastry according to the instructions on page 203.
+ Adjust the oven shelf to the middle position, preheat the oven to 200°C, line two baking sheets with baking paper and remove the rough puff pastry from the fridge or freezer to come up to room temperature as directed on page 203 (to prevent cracking).
+ Roll out the pastry to about 2mm thick and cut out a 23 x 50cm rectangle. Use the whisked egg to wash the pastry and scatter half the grated parmesan evenly across the top. Press down lightly with your palms to make sure the parmesan sticks.
+ Flip the pastry over, so that the cheesy side faces downwards. (If you find this size of pastry difficult to flip, you can divide it into 2 smaller 23 x 25cm rectangles and handle them independently.) Brush the other side with the whisked egg and sprinkle with the remaining parmesan.
+ Cut the dough into 8–10mm-wide and 23cm-long strips – you should get about 50. Take 2 strips and twist them around each other, making them lie as flat as possible. ①
+ Repeat with the remaining strips, to give 25 cheese twists. Transfer them to the lined baking sheets and refrigerate for about 15 minutes.
+ One baking sheet at a time, bake the twists for about 15–20 minutes or until puffed up and deep golden brown. ② Allow to cool on a wire rack, then serve.

STORAGE

Best on the day, but you can keep them in a closed container for up to 2 days.

RASPBERRY FRANGIPANE TART

SERVES 10–12
PREP TIME 45 mins
(excluding the pastry)
CHILL TIME 15 mins
COOK TIME 5 mins
BAKE TIME 45 mins

1 quantity of **sweet shortcrust pastry** (see page 204)

FRANGIPANE FILLING
150g unsalted butter, softened
200g caster sugar
1 teaspoon vanilla bean paste
2 eggs, at room temperature
1 egg yolk, at room temperature
200g almond flour
100g gluten-free flour blend

TO ASSEMBLE
200g raspberry jam
1 tablespoon lemon juice
150–200g raspberries
30g flaked almonds

NOTES
① *Mixing the frangipane on the lowest speed ensures minimal aeration (avoid using a hand mixer for the same reason). Trapped air would cause the filling to puff up during baking, resulting in a cakier texture. If you don't have a stand mixer, mix the filling by hand, using a wooden spoon or spatula.*

② *Reducing the raspberry jam removes much of the moisture, making it less likely to bubble up through the frangipane filling during baking.*

This take on a Bakewell tart has fresh raspberries dotted around the frangipane filling. The raspberries provide an inviting burst of colour, and their tartness balances out the sweetness of the frangipane. You don't have to blind bake the pastry shell – the relatively long baking time means the pastry bakes through to crisp, buttery perfection precisely when the filling reaches the point between gooey and moist.

FRANGIPANE FILLING
+ Using a stand mixer with the paddle attachment on the lowest speed setting, fully combine the butter, sugar and vanilla paste. ①
+ With the mixer still on its lowest setting, add the eggs, one at a time, then the yolk, mixing well after each addition, until smooth.
+ Add the almond flour and gluten-free flour blend, and mix until combined. Set aside until needed.

ASSEMBLING THE TART
+ Pre-make the pastry according to the instructions on page 204.
+ You'll need a 23cm loose-bottomed, fluted tart tin. Bring the pastry up to room temperature, then roll it out on a lightly floured surface to a circle about 3mm thick and about 4cm larger than the tin diameter. (If the pastry is slightly crumbly, knead it a little first.)
+ Transfer the pastry circle to the tin and make sure it's snug against the bottom and sides. Roll a rolling pin across the tart tin to trim away any excess, and gently press the pastry into the grooves of the fluted edge. Chill for at least 15 minutes. Adjust the oven shelf to the middle position and preheat the oven to 180°C with a large baking sheet inside.
+ In a saucepan, heat the raspberry jam and lemon juice until runny. Pass the mixture through a sieve into a bowl to remove any seeds, then return it to the saucepan and cook over a medium–high heat for about 5 minutes to reduce it slightly – it should be thick and viscous. ② Remove from the heat and allow to cool to room temperature.
+ Spread the cooled, thickened raspberry jam evenly across the bottom of the pastry. Spoon the frangipane on top of the jam and smooth out the top. Place the fresh raspberries on top of the frangipane, gently pressing them down into the filling. Sprinkle the flaked almonds evenly over the top.
+ Place the tart on the hot baking sheet and bake for 45–50 minutes or until golden brown and the filling feels set and doesn't wobble when you gently shake the tin. If the top starts browning too quickly (before the filling is baked), cover with foil (shiny side up) and bake until done.
+ Allow the tart to cool in the tin for about 15–20 minutes, then remove and transfer to a wire rack to cool completely.

STORAGE
3–4 days in a closed container at room temperature.

STRAWBERRIES + CREAM TART

SERVES 10–12
PREP TIME 1 hour 30 mins
(excluding the pastry)
CHILL TIME 8 hours 15 mins
BAKE TIME 2 hours 25 mins
COOK TIME 10 mins

1 quantity of **sweet shortcrust pastry** (see page 204)

VANILLA CREAM

325g whole milk
1 teaspoon vanilla bean paste
4 egg yolks
100g caster sugar
35g cornflour
35g unsalted butter
250g double or whipping cream, chilled
75g icing sugar, sifted

TO ASSEMBLE

500g strawberries, some whole, some hulled and thickly sliced
50g caster sugar
2 tablespoons lemon juice
300g double or whipping cream, chilled
100g icing sugar, sifted

MERINGUE KISSES (MAKES 30)

2 egg whites
75g caster sugar
¼ teaspoon cream of tartar

Two words: Summer. Showstopper. This tart is based on the classic British combination of strawberries and cream, with an Eton-mess twist. Aside from the mouthwatering flavours, the tart provides an exciting texture experience, with the creaminess of the vanilla and whipped creams contrasting beautifully with the crispness of the shortcrust pastry and the sweet crunch of meringue kisses.

TART SHELL

+ Pre-make the pastry according to the instructions on page 204.
+ You'll need a 23cm loose-bottomed tart tin with a fluted edge. Adjust the oven shelf to the middle position and preheat the oven to 180°C with a large baking tray or sheet inside.
+ Roll out the pastry on a lightly floured surface to a circle about 3mm thick. (If the pastry is slightly crumbly, give it a knead to make it more pliable.) The pastry should be about 4cm larger than the tin diameter (to account for the height of the tin sides).
+ Transfer the pastry into the tin, ensuring it's snug against the bottom and sides. Roll a rolling pin across the tart tin to trim away any excess and gently press the pastry into the grooves of the fluted edge, like you're crimping it. ① Refrigerate for at least 15 minutes.
+ Prick the chilled pastry base with a fork, line the inside with scrunched-up baking paper and fill with baking beans or rice.
+ Place the pastry case on the hot baking sheet and blind bake for 15 minutes, then remove the baking paper and baking beans or rice and bake for a further 10–15 minutes until the pastry is an even golden brown and shrinks away from the tin.
+ Allow the pastry to cool completely to room temperature in the tart tin and set aside until needed.

VANILLA CREAM

+ In a saucepan, cook the milk and vanilla paste over a medium–high heat until boiling. In the meantime, whisk together the egg yolks and sugar until pale and smooth, then add the cornflour and combine. ②
+ Pour the hot milk in a slow stream into the egg-yolk mixture, whisking continuously. ③ Return the mixture to the pan and cook over a high heat, whisking continuously, until thickened (about 1–2 minutes).
+ Remove from the heat and stir in the butter, mixing well until the butter has melted and the cream is smooth and glossy.
+ Allow the pastry cream to cool completely, stirring or whisking occasionally to prevent a skin forming. ④
+ Using a stand mixer with the whisk or a hand mixer fitted with the double beaters, whip together the cream and icing sugar to soft peaks.
+ Briefly whip the cooled pastry cream until smooth and glossy, then fold in the softened whipped cream until combined. Set aside until needed.

ASSEMBLING THE TART – PART I

+ Using a food processor or blender, purée 300g of the strawberries with the caster sugar and lemon juice until smooth. Pass the mixture through a sieve into a bowl to remove any seeds, then transfer to a pan.
+ Cook over a medium–high heat for about 5 minutes, stirring frequently, to reduce the juices and until slightly thickened and viscous. You're looking for a thick, runny consistency (you should be able to drizzle it off a spoon), but not as thick as jam. Allow to cool completely.
+ Drizzle the strawberry reduction into the vanilla pastry cream and fold it in with a couple of strokes to get a ripple effect.
+ Transfer the strawberry ripple into the tart shell, smooth out the top and refrigerate for at least 8 hours, or preferably overnight. While the tart is chilling, prepare the meringue kisses.

MERINGUE KISSES

+ Adjust two oven shelves to the middle and lower positions, preheat the oven to 100°C and line two large baking sheets with baking paper.
+ Mix the egg whites, sugar and cream of tartar in a heatproof bowl over a pan of simmering water. Stir frequently until the meringue mixture reaches 65°C and the sugar has dissolved. ⑤
+ Transfer the mixture to the bowl of a stand mixer with the whisk attachment, or use a hand mixer fitted with the double beaters, and whisk on medium–high speed for 5–7 minutes to stiff, glossy peaks.
+ Transfer to a piping bag fitted with a large star nozzle, and pipe about 3.5–4cm meringue stars on to the lined baking sheets, leaving about 1cm between them (you'll have about 30 kisses – you can store any leftovers in an airtight container for up to 2 weeks). Alternatively, pipe 7.5cm meringue nests or 1cm meringue stars, if you prefer.
+ Bake the kisses for about 2–2½ hours, or until you can lift the meringue kisses off the baking paper without them sticking or falling apart. The meringues should sound hollow when you tap them.
+ Allow to cool in the oven with the door closed, then remove from the oven and store in a closed container until needed.

ASSEMBLING THE TART – PART II

+ Remove the tart from the tart tin and decorate it just before you intend to serve (the meringue kisses soften quickly once they are in contact with the moisture from the cream and strawberries). Whip the cream with the icing sugar until soft peaks form.
+ Dollop the whipped cream on top of the chilled tart, and decorate with the remaining strawberries and the meringue kisses. Serve immediately.

NOTES

① *Pressing the pastry into the grooves of the tart tin is a small detail, but that extra few minutes' work gives a much more polished finish.*

② *Whisking the egg yolks and sugar together until smooth is called 'blanching'. The sugar protects the egg proteins, preventing lump formation during cooking.*

③ *This is called tempering, and it prevents the egg yolks from scrambling, which would happen if you added the mixture all at once to the hot milk. Tempering slowly increases the temperature of the egg yolks while at the same time diluting them, to give a silky-smooth pastry cream.*

④ *You could follow the more traditional method of covering the surface of the pastry cream with cling film. However, I find that whisking is sufficient to prevent the skin forming, and you don't waste the cream that would get stuck to the cling film.*

⑤ *Using Swiss meringue for the meringue kisses eliminates any concerns about the egg whites not being heated or cooked properly. It is very stable, pipes beautifully and dries to give perfectly crunchy meringue kisses that melt on your tongue.*

PHOTOGRAPHY OVERLEAF ⟶

HOT CHOCOLATE TART

SERVES 10–12
PREP TIME 1 hour
(excluding the pastry)
CHILL TIME 4 hours 15 mins
BAKE TIME 20 mins

Hot chocolate meets a fancy tart in this rather easy, yet definitely dressed-to-impress dessert. The blind-baked tart shell is filled with a chocolate mixture that lies in that delicious region between a chocolate mousse and a ganache, and everything is topped off with toasted Swiss meringue – because, after all, it would hardly be a proper hot chocolate (tart) without marshmallows.

1 quantity of **chocolate
   sweet shortcrust pastry**
   (see page 207)

CHOCOLATE FILLING
300g dark chocolate (about
   60% cocoa solids), chopped
400g double cream, chilled

SWISS MERINGUE
3 egg whites
150g caster sugar
¼ teaspoon cream of tartar

TART SHELL
+ Pre-make the pastry according to the instructions on page 207.
+ You'll need a 23cm loose-bottomed tart tin with a fluted edge. Adjust the oven shelf to the middle position and preheat the oven to 180°C with a large baking tray or sheet inside.
+ Remove the chocolate shortcrust pastry from the fridge and bring it up to room temperature as directed on page 207. Briefly knead the pastry to even out its texture and make it more flexible.
+ Roll out the pastry on a lightly floured surface to a circle about 3mm thick. The pastry should be about 4cm larger than the tin diameter, to account for the height of the tin sides.
+ Transfer the pastry into the tin and make sure it's snug against the bottom and sides. Roll a rolling pin across the top of the tart tin to trim away any excess, and gently press the pastry into the grooves of the fluted edge, like you're crimping it. ① Refrigerate for at least 15 minutes.
+ Prick the chilled pastry base with a fork, line the inside of the case with scrunched-up baking paper and fill it with baking beans or rice.
+ Place the the pastry case on the hot baking sheet and bake for 15 minutes, then remove the baking paper and baking beans or rice and bake uncovered for 5–10 minutes, until the pastry feels firm and crisp to the touch and shrinks away from the edges of the tin.
+ Allow the pastry to cool completely to room temperature in the tart tin and set aside until needed.

CHOCOLATE FILLING

+ Place the chopped dark chocolate into a heatproof bowl.
+ Pour 240g of the double cream into a saucepan and bring to a simmer. Pour the cream over the chocolate. ② Allow to stand for 3–4 minutes, then stir until you get a smooth and glossy ganache. Allow to cool until warm.
+ In a separate bowl, whip the remaining 160g of double cream until stiff peaks only just begin to form. ③
+ Gently fold the whipped cream into the cooled ganache until fully incorporated.
+ Transfer the chocolate filling into the cooled tart shell and smooth out the top, creating swirls with an offset spatula or the back of a spoon.
+ Refrigerate the filled tart for at least 4 hours or until the filling has completely set and firmed up.

SWISS MERINGUE

+ Prepare the meringue topping as close to serving as possible.
+ Mix the egg whites, sugar and cream of tartar in a heatproof bowl over a pan of simmering water. Stir continuously until the meringue mixture reaches 65°C and the sugar has dissolved.
+ Remove from the heat, transfer to a stand mixer fitted with the whisk attachment, or use a hand mixer fitted with the double beaters, and whisk on medium–high speed for 5–7 minutes until greatly increased in volume and stiff peaks form. Make sure the meringue is at room temperature before you spread it on top of the chocolate mousse filling.
+ Remove the chilled chocolate tart from the tart tin, top it with the Swiss meringue and toast it using a kitchen blow torch. Then, slice the tart with a sharp knife dipped in hot water, to serve.

STORAGE

3–4 days in a closed container in the fridge.

NOTES

① *Pressing the pastry into the grooves of the tart tin is a small detail, but that extra few minutes' work gives a much more polished finish.*

② *You can tweak the relative amounts of the double cream used in the ganache and the double cream whipped up to stiff peaks (keeping the total weight constant) to fine-tune the texture of the chocolate filling. Increasing the amount used in the ganache makes the filling denser and truffle-like, while increasing the amount of the whipped cream gives a fluffier, lighter, mousse-like result.*

③ *Because the filling uses a 60% cocoa-solids chocolate, I find it sweet enough as it is. However, if you want to increase the sweetness, whip the double cream together with about 50–100g icing sugar, before folding it into the ganache.*

PHOTOGRAPHY OVERLEAF ⟶

BREAD

Bread. The holy grail of gluten-free baking.

If I asked you here and now to tell me whether you think it's possible to make gluten-free bread that tastes and feels and smells like regular bread, what would you answer? (No cheating by flicking through the next few pages.)

Probably, you'd answer with a very sceptical (and very confident), 'No.' I'm here to tell you that you'd be wrong. In this chapter you'll find gluten-free bread recipes that handle like regular bread dough, go through two rounds of rising, and have a crisp, caramelised crust and a soft, chewy crumb. Just. Like. Regular. Bread.

Now, don't get me wrong. If you compare regular and gluten-free bread recipes, you'll find many (very important) differences – the main one being that regular bread relies on prolonged kneading that 'develops' the gluten and gives the dough its characteristic elasticity and extensibility, from which everything else follows. With gluten-free bread that's not possible. Obviously.

But here's the important bit: with just a few handy tricks up your sleeve, you can replicate all your favourite carby delicacies, even in the absence of gluten. From crusty artisan loaves and soft sandwich bread, all the way to decadent cinnamon rolls and indulgent fried doughnuts – it's all possible and achievable with the help of the recipes and the science within this chapter. And the very best part? It's not difficult at all.

THE TYPES OF BREAD IN THIS CHAPTER

Much like with the rest of the book, when it comes to bread, I really want you to get a sense of what's possible in the world of gluten-free baking. With that in mind, this chapter covers:

+ big loaves, baked in baking tins or 'free-form'
+ buns and rolls
+ other doughs, like flat bread and pizza dough
+ enriched (brioche-type) doughs

My promise to you is this: once you've mastered the basic recipes, there will be no limit to the gluten-free bread you'll be able to bake. Once you understand the science, and know the reasons for the different ingredients and methods, you'll be able to convert pretty much any regular bread recipe into one that is gluten-free. Doesn't that sound like an absolute dream?

THE ROLE OF GLUTEN IN WHEAT-BASED BREAD

Before we enter the exciting world of gluten-free bread baking, let's take a look at the role that gluten plays in wheat-based bread, so that we understand what we're trying to mimic (and why). In wheat bread:

+ Gluten provides structure through the formation of an intricate interwoven network of protein strands, resulting in the characteristic chewy crumb.

+ This network helps to capture the gases produced by the yeast action, which increases the volume of the bread – also known as rising and proving.

+ Through the same network, gluten provides a certain surface tension when you shape the dough, which helps to control the shape of the bread.

+ And, finally, because of its chemical composition (it comprises two components: glutenin and gliadin, both of which can be hydrated and can hold on to water), gluten helps with moisture retention and so prevents the bread from drying out too quickly.

At this point, it should be obvious that gluten is pretty essential in bread baking. However, this fact by no means poses an insurmountable obstacle – with some clever additions and a few tweaks, you'll be able to get perfectly baked bread even in the absence of gluten.

THE IDEAL LOAF OF BREAD

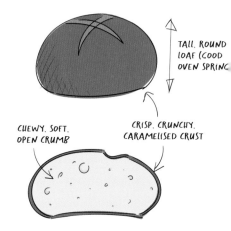

TALL, ROUND LOAF (GOOD OVEN SPRING)

CHEWY, SOFT, OPEN CRUMB

CRISP, CRUNCHY, CARAMELISED CRUST

BEFORE WE GO ON: BAKER'S PERCENTAGE

In this chapter, you'll see that the recipes include information about the composition of the dough (such as the starch content and hydration) in terms of baker's percentage. As a shorthand, you'll see it abbreviated to b%, so as to differentiate it from regular percentage (%).

Baker's percentage is a useful way to express the proportion of ingredients (or groups of ingredients) in a bread recipe. This proportion stays constant even if you scale the recipe up or down. In using baker's percentage, the amount of each ingredient is expressed as a percentage of the flour weight (the flour weight is always 100 b%). That is:

ingredient b% = (weight of ingredient/weight of flour) x 100%

Baker's percentage is incredibly useful when it comes to comparing different recipes or converting regular bread recipes into gluten-free ones. As you'll see shortly, the chemistry (and therefore the success) of gluten-free bread baking is all about ingredient ratios, and baker's percentage is a simple way to keep track of this.

THE BINDERS: PSYLLIUM HUSK + XANTHAN GUM

While most recipes in this book contain only xanthan gum as the gluten substitute, the bread recipes contain both xanthan gum and psyllium husk. On the surface of it, the two binders carry out similar roles in gluten-free baking, but in practice they are very different, both in terms of how they affect the handling of the dough, and in terms of the properties they impart to the baked bread.

Psyllium husk
Where bread is concerned, psyllium husk is undoubtedly the more important of the two binders. It is psyllium husk that makes it possible to knead and shape the dough and creates a dough texture (its elasticity and 'sponginess', if you will) resembling that of wheat bread. Without psyllium husk, you are limited to pourable or scoopable

By absorbing water (or other liquids) and forming a gel, psyllium husk transforms what would otherwise be a runny, pourable bread 'batter' into a dough you can knead and shape whichever way you want.

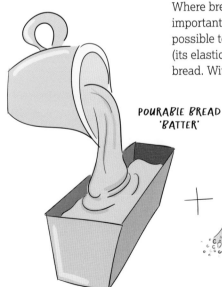

POURABLE BREAD 'BATTER'

PSYLLIUM HUSK

GLUTEN-FREE BREAD DOUGH YOU CAN KNEAD & SHAPE

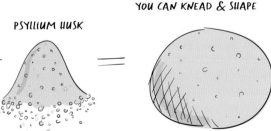

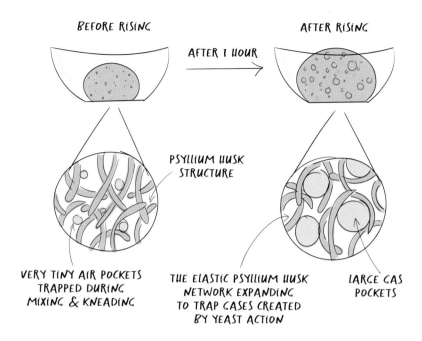

BEFORE RISING

AFTER RISING

AFTER 1 HOUR

PSYLLIUM HUSK STRUCTURE

VERY TINY AIR POCKETS TRAPPED DURING MIXING & KNEADING

THE ELASTIC PSYLLIUM HUSK NETWORK EXPANDING TO TRAP GASES CREATED BY YEAST ACTION

LARGE GAS POCKETS

Psyllium husk mimics gluten in that it creates an elastic network that helps trap gas bubbles produced by the yeast action. This, in turn, allows the dough to rise and double in volume – just like wheat bread.

bread 'batters' that more closely resemble cake batter than bread dough. Add psyllium husk, and all of a sudden you transform the sticky, uncontrollable mess into a dough that handles like a dream.

When mixed with water (or other liquids), psyllium husk forms a slightly sticky, elastic gel that acts as the gluten substitute – before, during and after baking. By mimicking gluten, psyllium husk creates the flexibility and structure the dough needs to trap the gases the yeast produces and expand with them during the two rounds of rising. It also gives the bread that authentic chewy crumb we all know and love.

You can buy psyllium husk in two forms: as the rough husk (also known as 'whole' psyllium husk) or as a fine powder. All my recipes refer to the rough husk (as it's the one I prefer to use). If you have only psyllium husk powder to hand, you will need to use a slightly smaller quantity than the one listed in the recipe. That's because the powder absorbs more liquid than the rough husk and therefore forms a stiffer gel. As a good rule of thumb, for every 1g of whole psyllium husk listed in the recipe, use 0.85g of psyllium husk powder.

Xanthan gum

If psyllium husk is responsible for the chewiness of the bread, then xanthan gum adds the pillowy softness we associate with a loaf of white bread or brioche-type sweet dough. Xanthan gum also helps to create the characteristic 'flake' of regular bread. We don't normally think of bread as 'flaky', but think back to wheat-based brioche buns or cinnamon rolls baked closely together. When you tear them apart, the bread flakes at the juncture points. In gluten-free baking, you don't get that by using psyllium husk only. Instead, you need a mixture of the two binders.

We don't typically think of bread as 'flaky' – but wheat-based brioche buns (as well as cinnamon rolls and pull-apart rolls) flake at juncture points where they're baked closely together and join as they expand in the heat of the oven.

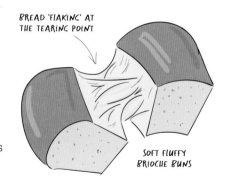

BREAD 'FLAKING' AT THE TEARING POINT

SOFT FLUFFY BRIOCHE BUNS

Experiment with sticky caution

Finally, when you start experimenting on your own: don't be tempted to go overboard with the amount of binder you add. While it's true that too little binder leads to a loose, unmanageable dough, too much causes just as many (if different) problems: a stiff dough that is almost impossible to shape without cracking and that struggles to expand and rise. In general, 4–6 b% of binders is more than enough to achieve gluten-free-bread perfection.

PROTEIN VS STARCH GLUTEN-FREE FLOURS

I touched on the distinction between starch and protein gluten-free flours in the previous chapters, but while it was of passing interest in making cakes, brownies and cookies, it's at the very forefront of the science of gluten-free bread.

Protein gluten-free flours

Broadly speaking, protein gluten-free flours are the hearty, flavour-carrying components of a bread-flour mixture. They also contribute to the bread's elasticity and structure (albeit significantly less so than regular wheat flour), and their higher moisture retention means that gluten-free bread with a high protein flour content dries out much more slowly. You can keep such bread in a closed container (or even just wrapped in a tea towel) for four or even up to five days, without it becoming dry or crumbly.

Starch gluten-free flours

On the other hand, starch gluten-free flours give bread a much softer texture and an open, airier crumb. When it comes to pillowy burger buns or rich, soft cinnamon rolls, starch flours are your best friends.

However, I don't recommend using just one flour type in a bread recipe. Bread made exclusively from protein flours can be dense and heavy, while one made only from starch flours can have little-to-no flavour and dries out very quickly. Instead, it's all about the relative quantities of protein and starch flours. To eliminate any guesswork (at least initially, while you're still getting used to the basics), I've provided the starch content of the bread, in terms of baker's percentage, at the end of each recipe.

The importance of finely ground flours

Finally, it's important that the flours you use are finely milled (sometimes known as 'superfine'). They should have a powdery texture – think regular wheat flour rather than polenta. A coarse flour is more difficult to hydrate (that is, it absorbs less liquid) and can consequently lead to disappointing results.

GLUTEN-FREE FLOUR SUBSTITUTIONS

I'm aware that you might have certain sensitivities, intolerances or allergies that may, for example, prevent you from using cornflour, potato starch or rice flours. Broadly speaking, you can swap most starch gluten-free flours for one another with little or no noticeable effect on the taste or texture of the bread (with the exception of white rice flour, which is a bit of an outlier in the starch family – see page 16).

For protein flours, the situation is somewhat more complicated. This is less owing to the different protein and elasticity properties of the flours, and more because of the variable amount of heaviness and flavour they impart. In general, protein flours can be subdivided into two groups, which contain interchangeable flours:

+ **brown rice and millet flours** (on the 'lighter' end of the heaviness spectrum and milder in flavour),
+ **buckwheat, maize, oat, quinoa, sorghum and white teff flours** (on the 'heavier' end of the spectrum and more intense in flavour).

While I encourage you to experiment with different flours to find your personal favourites, unless you have an allergy, I do recommend you start by using the flours specified in the recipes. I have chosen these particular flours and their relative proportions because they give the best results in terms of texture, flavour and appearance.

THE YEAST

I use active dried yeast, but you can use instant (quick-rise) yeast, if you prefer. You'll need only 0.75g instant yeast for every 1g active dried yeast – mainly owing to how the two yeasts are processed. (To use fresh yeast, you'll need double the weight of active dried yeast.)

The reason I like active dried yeast lies in the way we use it. It needs to be 'activated' in a warm liquid, generally water or milk – I usually also add a tablespoon or so of sugar to really kick-start the yeast action. Within 10 minutes, the yeast mixture will get frothy and bubbly, a clear indication that the yeast is active. If nothing happens and no froth appears, you know that the yeast is dead and you need a new batch.

Instant yeast, on the other hand, doesn't need activating. You can add it directly to the dough along with the dry ingredients. You then add separately the water that you would have used activating the yeast, along with the psyllium gel. Although this is more efficient, you're skipping the step that tells you whether your yeast is active (if it's not, your bread is doomed). Practicality aside, I find that there's something magical about seeing the yeast bubble away before you start, almost like the mixture is telling you that it's ready to get down to business and that good things are coming your way.

THE HYDRATION

The one thing you should take away from this section is that, on average, the hydration of gluten-free bread is much higher than that of wheat bread. If you look at a range of wheat bread recipes, you'll see hydration percentages in the range of about 60 to 80 b%. (Note that hydration is often expressed in terms of 'normal' percentages, but we're going to stick with baker's percentage for consistency's sake.) However, the non-enriched gluten-free bread recipes (that is, those that don't contain eggs and/or butter) in this chapter usually have hydrations higher than 100 b% – sometimes as high as 118 b%!

That's because gluten-free flours have a much higher water-absorption capacity than wheat flour, meaning that if you were to use the same amount of water as in a regular bread, the gluten-free bread would crumble away to nothing. Furthermore, psyllium husk (and to a smaller extent xanthan gum) also absorbs a significant amount of water, raising the required hydration even higher.

So, please don't panic or doubt the recipe when it tells you that you need a lot of water to make this or that gluten-free bread. You do and the recipe is right. A bread with a too-low hydration is too stiff to rise properly, tends to crack as its volume increases, and ends up having a dense, under-proved, dry and crumbly texture.

That said, when it comes to enriched breads that contain eggs, your dough will need a much lower hydration. (Note that I don't include the eggs in the hydration b%.) That's because psyllium husk interacts only negligibly with the moisture contained within eggs – so while psyllium husk readily absorbs water and, to a smaller extent, milk to give a manageable dough, the eggs act to make the dough softer and stickier. If, on top of that, the hydration was as high as for non-enriched gluten-free dough, you'd end up with a dough soup and no hope of making something edible out of it. The hydration of enriched dough therefore usually doesn't exceed 70–85 b%.

HANDLING + KNEADING THE DOUGH

You can knead all the breads and doughs in this book either by hand or using a stand mixer fitted with the dough hook. Although using a stand mixer is definitely much more convenient, I recommend that when you begin baking gluten-free bread at home, you mix and knead by hand. Get in there, squeeze the dough with your fingers, and feel its texture and consistency. As well as being incredibly relaxing, this will help you develop a certain intuition when it comes to making gluten-free bread. Sure, the science and the baker's percentage calculations should still take centre stage, but actually feeling the dough will allow you to – in time – make minor adjustments such as adding an additional splash of water or a handful of flour just based on how the dough feels.

There's really no right or wrong way to knead gluten-free bread, as long as you end up with a smooth dough throughout which all ingredients are homogeneously dispersed. Once you have combined the dry and the wet ingredients, the easiest way to knead the dough by hand is to squeeze it through your fingers, going around the bowl and scraping any patches of flour off the sides and bottom. As you're not trying to develop the gluten and therefore the elasticity of the dough, as you would with wheat bread, there's no need for the stretching motions we typically associate with kneading.

Once the dough is smooth, I like to shape it into a ball before its first rise. For the best results, do this on a lightly oiled surface with lightly oiled hands (use a neutral-tasting oil, such as vegetable or sunflower oil). However tempting, avoid dusting your hands and the work surface with flour at this stage – incorporating extra flour into the dough will alter its properties, deviating from the perfect flour and hydration profiles.

FIRST RISE: BULK FERMENTATION

The first rise is more correctly described as bulk fermentation and it's during this that the flavour develops. At this stage, enzymes called amylases degrade the starches in the flour down to simple sugars. The yeast then consumes these sugars, releasing a gas (carbon dioxide), which is trapped within the bread, causing it to expand. At the same time, the sugars are converted into alcohols, which help to give bread its delicious flavour.

I usually leave my breads to rise at temperatures of around 26–32°C, which is a sweet spot for high yeast activity. In the hot summer months that means leaving the bread to rise on the kitchen counter at room temperature. When the weather is colder, you can place it on top of a warm radiator, or in a warm oven. If your oven has several compartments and you're using one to bake a cake or cookies,

In the absence of a proving drawer, my go-to rising and proving spot for my loaves is a warm oven, or the adjacent compartment of an oven that's baking some other irresistible treat at the same time.

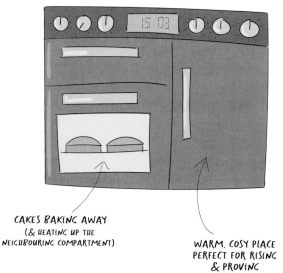

CAKES BAKING AWAY
(& HEATING UP THE
NEIGHBOURING COMPARTMENT)

WARM, COSY PLACE
PERFECT FOR RISING
& PROVING

If you're not confident about recognising when your dough has doubled in volume, use a see-through container with vertical walls. Position a rubber band or a piece of string around the container to mark the starting height of the dough – when it's twice that height, the volume has doubled, too.

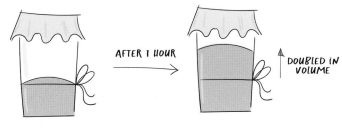

MUCH EASIER TO SEE THAN:

the neighbouring compartments will be the perfect warm, cosy place for the dough to rise. Alternatively, you can heat up your oven until just warm, turn it off and let the dough rise in there, making sure not to open the oven too often. All that said, don't worry if your chosen rising spot is a bit cooler than ideal: your dough will take a bit longer to double in volume, but it will be just as delicious in the end.

If eyeballing volumes isn't your strong point (and I'm speaking from experience here), use a see-through container with vertical walls, such as a plastic storage container, for rising the dough. Then, just mark the initial volume with a rubber band or a piece of string or tape and wait for the dough to reach double that height.

SHAPING THE DOUGH

Once doubled in volume, the dough needs shaping before its second rise. For best results, do this on a lightly floured surface – but don't be tempted to add more flour into the dough, even if it feels soft and slightly sticky to the touch. While adding more flour might make the dough more manageable, it's also likely to result in a dense, dry crumb.

I recommend shaping enriched doughs, such as cinnamon rolls or chocolate babka (which need to be rolled out, filled and then rolled up), on a lightly floured sheet of baking paper. This way, if you need to refrigerate the dough, you can easily slide it on to a baking sheet and into the fridge. Furthermore, when rolling the dough into a log, the baking paper can act as a prop: just lift it up on one side and allow gravity to do its thing to roll it nice and tight.

Even with perfect shaping (be that shaping large loaves into a ball or rolling out baguettes), you'll notice that compared to wheat-based dough, gluten-free breads lack some of the surface tension that's responsible for the smooth surface of a regular burger bun or a loaf of sandwich bread. That's purely down to the absence of the gluten network, which even the perfect amount of binders can't

100% replicate. That said, with correct shaping the difference is only minor, and certainly doesn't take away from the deliciousness.

THE SECOND RISE: PROVING

In addition to further fermentation and therefore flavour development, the second rise (the prove) is responsible for the bread's final fluffiness. This can happen in a baking tray or tin (the container you intend to bake the bread in), in a proving basket (for breads that you intend to bake in a preheated skillet or Dutch oven) or 'free-form' (for more rustic loaves with a less well-defined shape or for smaller buns and rolls).

Of course, it's impossible to talk about the prove without mentioning the dreaded notion of over-proving. In addition to the yeast converting sugars into alcohols and releasing carbon dioxide along the way, enzymes present in the dough are working away, breaking down the starches. Leaving the dough to prove for too long can compromise its structure – it will over-prove and is likely to collapse during baking.

In one sense, over-proving is a scary subject, especially in the realm of gluten-free bread where the absence of gluten makes the dough even more fragile. In another, though, being a scaredy-cat and baking the bread too early in fear of a collapsing loaf can lead to under-proved bread and a tight, unpleasant crumb.

Judging when the dough is perfectly proved is trickier for gluten-free bread than for wheat-based bread. The frequently used 'poke test' (where you poke your dough and see if it springs back) isn't applicable in gluten-free loaves as it relies strongly on the presence of gluten. Plus, it could actually deflate the bread, destroying all the effort you've put in so far.

The best guide for whether or not gluten-free bread is ready for baking is its volume: look for a dough that has increased in volume by at least 50% and at the most doubled in volume. Beyond that, it's a question of experience. After you've baked a particular loaf successfully a couple of times, you'll know how far it should rise in a particular baking tin or proving basket before it's ready for the oven. To help you, all my recipes list approximate proving times.

BAKING THE BREAD

Baking the proved dough involves more than just popping it into the oven, closing the door and hoping for the best. You might say that's unfortunate – but actually, it gives us even more opportunity to control and improve on the final outcome.

There are six stages of baking delicious gluten-free bread:
1. preheating the oven
2. glazing the dough (if required)
3. scoring the dough
4. oven spring and introducing steam into the oven (if required)
5. Maillard reaction and caramelisation
6. drying out the crumb

The list might seem somewhat overwhelming, especially considering how much information I've (hopefully) unloaded into your brain already. But it would be a terrible shame if you put all this effort into the right ingredients, getting your hands sticky with kneading and waiting for two rounds of rising only to falter when gluten-free-bread perfection is in sight. So, read on. It's worth it.

1. Preheating the oven
Depending on the type of gluten-free bread, the recipe will call for an oven temperature somewhere in the 180–250°C range. Because you want everything to be ready to go when your bread hits the point of being perfectly proved, start preheating the oven at least 30–45 minutes before you expect the dough to be ready. This will ensure that the oven is at a stable temperature and that any skillets, Dutch ovens or water trays that also need preheating have had sufficient time to do so.

2. Glazing (optional)
This is mostly an aesthetic consideration but, as we also eat with our eyes, one worth mentioning. The surface of gluten-free bread likes to dry out in the oven, sometimes even in the presence of a sufficient amount of steam. Consequently, it can develop an unsightly white-ish crust that, while still crisp and tasty, is a far cry from deep golden-brown deliciousness. This isn't a consideration for all breads, and I've made a note as relevant in individual recipes. (Also, this is different from brushing the finished bakes with butter or a sugar syrup after baking, both of which help keep the crust softer and the overall flavour more delicious.)

Although there are many possible glazes (whole eggs, egg yolks only, egg whites only, water, milk, cream, butter, oil – and combinations thereof), I usually stick to the following:

+ **whisked whole egg** for a deep golden-brown crust with a nice shine to it,

+ **oil or melted butter** for a deep golden-brown crust with a matte-looking finish.

3. Scoring the dough

There are two reasons for scoring the dough: it makes for prettier loaves (especially if you score in an intricate pattern) and it controls the way in which the bread expands in the oven, resulting in a better final crumb. Note that this is relevant only for crusty breads, such as artisan loaves, and not for enriched doughs like cinnamon rolls or babka.

As the bread encounters the very high heat of the oven, the yeast gets a last big boost in activity and generates even more gas. Additionally, some of the water in the dough transforms into steam, which rapidly expands owing to the high oven temperature. These factors combine to create a process called 'oven spring' (more on that below). In the absence of scoring, this can result in random cracks along the surface of the bread and a misshapen final loaf or, and this is especially true for gluten-free bread, in a bread that despite the heat expands only negligibly and therefore ends up having a much tighter crumb.

4. Oven spring + introducing steam

As the bread encounters the high temperatures within the oven, two things happen. First, before the high heat kills the yeast (which survives up to a maximum temperature of about 55–60°C), it generates one last burst of carbon dioxide. Second, the new carbon dioxide along with the gases already trapped in the bread (as well as part of the water and the alcohols produced by fermentation that have evaporated) undergo a rapid expansion. This, in turn, causes the bread to expand along with them – this is oven spring, a process that takes place roughly during the first 10 minutes of baking.

We've talked about the importance of scoring to control how bread expands – but there's a second factor to consider: the rate at which the crust dries out and firms up. As you might expect, a crust that becomes crisp and hard the moment the bread enters the oven is counterproductive to oven spring. This is mostly a consideration for crusty bakes, such as artisan loaves or baguettes. In comparison, enriched doughs have a softer crust that crisps up much more slowly.

Oven spring refers to the bread rapidly increasing in volume in the first 10 minutes or so in the oven. Additional steam sources, such as ice cubes, keep the crust supple to allow the loaf to expand.

ICE CUBES (ADDITIONAL STEAM SOURCE)

VERY HOT CAST-IRON SKILLET/DUTCH OVEN

OVEN SPRING

HOT OVEN @ 250°C WITH STEAM

That's where steam comes in. Steam keeps the crust supple and malleable enough for the bread to expand, which helps to achieve an open crumb with plenty of holes. You can introduce steam into the oven using a couple of different methods (and you'll often see them combined for best results):

+ from the bread itself, a fact that is exploited in a Dutch oven or a combo cooker (where the closed container prevents the steam escaping),
+ by adding ice cubes – either into a hot tray at the bottom of the oven, or right next to the bread if it's baked in a Dutch oven, a combo cooker or a cast-iron skillet,
+ by adding boiling-hot water into a hot tray at the bottom of the oven,
+ by spraying the oven and/or the bread with water.

5. Maillard reaction + caramelisation

After the initial 10–20 minutes when oven spring has taken place and all the sources of steam have gone, the crust starts crisping up and getting a nice, deep golden-brown tan. What's occurring is actually really similar to the reactions that happen when you sear a piece of steak, torch marshmallows or even toast coffee beans.

This food-browning, accompanied by the formation of all sorts of delicious flavours, is known as the Maillard reaction, which occurs between amino acids and reducing sugars in the presence of heat. The amino acids are the result of broken-down proteins in the gluten-free flours, and the simple sugars are similarly derived from starch. The sugars also undergo a caramelisation process, which gives the crust an extra depth of flavour.

6. Crumb setting + drying

Even after the crust is golden brown, the bread-baking journey isn't quite over yet. First, the bread needs to reach an internal temperature of about 90–95°C so that it's baked all the way through and the crumb setting is complete. In simple terms, crumb setting is the transition of the bread interior from a sticky gloop to a porous, spongy structure we typically associate with perfectly baked bread. Second, the bread needs to achieve a sufficient level of water evaporation, otherwise the final crumb can be unpleasantly sticky and even collapse slightly on cooling.

WEIGHT (MOISTURE) LOSS DURING BAKING

Now for one of the most important questions in baking gluten-free bread: how do you know when the bread is baked? Conventional wisdom would tell you to do at least one of the following things:

+ **take its temperature** – the internal temperature should be about 90–95°C,
+ **listen to it** – when you tap it on the bottom, it should sound hollow,
+ **look at it** – perfectly baked bread should have a deep golden-brown crust.

However, none of these methods accounts for the fact that gluten-free breads need to lose a certain amount of water during baking in order to have a pleasantly chewy end crumb. Because gluten-free flours have a higher water-absorption capacity, gluten-free bread requires a longer baking time than wheat bread to achieve a sufficient level of water evaporation.

A loaf can reach the right internal temperature, it can sound hollow and it might have the most glorious dark golden crust – and yet, it might also need another 10–15 minutes in the oven to reach perfection. That's why I'm adding a fourth and crucial way of determining when your gluten-free bread is baked:

+ **weigh it** – and calculate the weight loss compared to the sum of all ingredients (that is, the weight of the bread before baking).

This might sound terribly tedious, but it's also one of the things that can mean the difference between the most glorious loaf of your life, and a wet, sticky disappointment (hidden behind a crisp, golden crust). And anyway, once you've made a particular bread a couple of times, you will already know what sort of final weight to expect, so you won't need to work it out every single time.

In very general terms, gluten-free bread should (and will, if baked correctly) lose anywhere from 12 to 24% of its weight during baking. To make your life easier, I've included information about the approximate weight of the bread after baking and the expected percentage weight loss at the end of each recipe.

There are, of course, a few exceptions – that is, breads and doughs where weighing doesn't produce any reliable or sensible results. The fillings and toppings in cinnamon rolls, chocolate babka and pizza render any weighing and weight-loss calculations meaningless. Similarly, the fact that we boil bagels and fry doughnuts makes any such weight-loss determinations pretty much impossible. But fear not – there are other ways to tell whether these bakes are done. The recipes tell all.

To determine whether your loaf of gluten-free bread is baked, you can take its temperature, listen to it, look at it and (most importantly) weigh it to determine the total moisture loss during baking.

THE SIGNS OF PERFECTLY BAKED BREAD

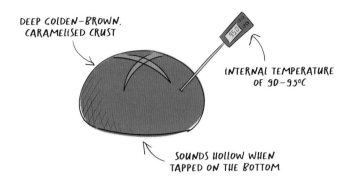

DEEP GOLDEN-BROWN, CARAMELISED CRUST

INTERNAL TEMPERATURE OF 90-95°C

SOUNDS HOLLOW WHEN TAPPED ON THE BOTTOM

BUT MORE IMPORTANTLY...

WEIGHT LOSS OF AROUND 12-24% DEPENDING ON BREAD TYPE

COOLING THE BREAD

Let me start by saying: do as I say, not as I do. I am probably the world's worst person when it comes to waiting for the bread to cool completely before cutting into it. But cooling the bread fully before tucking in is actually very important. Now, if I could only find enough patience to actually do that.

If you cut into hot bread, likelihood is your knife will be covered in sticky gloop and you'll think you're a big ol' gluten-free bread-baking failure. (Don't worry, you're really not!) This happens because there's still plenty of water vapour trapped in that hot loaf – as it cools, the water vapour will gradually evaporate and the bread will dry out. This additional moisture loss is actually fairly noticeable and can amount to a loss of up to 10–15g (equivalent to a 2–4% weight loss for most loaves in this chapter). Cooling also finishes the process that crumb setting started – any starches that haven't set previously will now 're-crystallise' into that familiar chewy, open bread texture.

Cooling the average loaf will take anywhere from one to two hours. Even if the bread feels cool to the touch, likelihood is its centre is still fairly warm. Don't stumble at this last hurdle. Be patient.

STORING GLUTEN-FREE BREAD

For all the talk about high water-absorption capacity, gluten-free bread also has a higher tendency to dry out in the days after it's baked. This becomes more noticeable the higher its starch content.

In general, store your gluten-free bread in a closed container, such as a bread box or plastic food container. For the most part, the bread will be fine this way for about three or four days, depending, of course, on the type of bread and the size of the loaf, among other things.

You can easily bring bread that's slightly dried out back to its original fluffy glory by reheating it briefly in the microwave in 5–10-second bursts or by toasting it in a pan on a bit of butter or oil.

THE TANGZHONG METHOD

Thought to have originated in Japan and popularised by Yvonne Chen, the author of *65°C Bread Doctor*, the tangzhong method uses a water or milk roux to make bread extra soft, tender and pillowy, as well as to prolong its shelf life. It's used mainly in milk and brioche-type doughs, such as the sandwich bread on page 276 and the burger buns on page 286.

Making the roux is incredibly simple – just whisk some of the particular gluten-free flour mixture (with xanthan gum already added) with the liquid of choice (normally water, milk or a mixture of the two). Then, cook it over a medium–high heat with continuous whisking until thickened into a pudding-like slurry (this happens when the roux reaches a temperature of about 65°C). Cool the roux until warm, then add it to the dough along with the rest of the wet ingredients.

By cooking the flour with the water, the starches in the flour gelatinise and the resulting gel locks in the additional moisture, which persists throughout the handling and the baking process – making the dough less sticky and the bread softer and moister.

Beginning at the bottom left, follow the arrows up and over for a summary of how to bake the best loaf of gluten-free bread of your life – starting at the ingredients and ending with a perfect slice of bread.

TURN OUT. SCORE & TRANSFER
INTO HOT CAST-IRON
SKILLET/DUTCH OVEN

2ND RISE
(FINAL PROVE)   26-32°C

OVEN SPRING

SHAPE THE DOUGH
& PLACE IN   26-32°C
PROVING BASKET

250°C

1ST RISE
(BULK FERMENTATION)   26-32°C

230°C   MAILLARD REACTION.
               CRUMB SETTING &
               DRYING

MIX & KNEAD
INGREDIENTS INTO
SMOOTH DOUGH   COOLING & FINAL
                      CRUMB SETTING

PATIENCE... LOTS OF IT

ARTISAN WHITE CRUSTY LOAF

MAKES 1 loaf
PREP TIME 45 mins
RISE TIME 2 hours
BAKE TIME 1 hour

8g active dried yeast
20g caster sugar
370g warm water
15g psyllium husk ①
125g tapioca starch
120g brown rice flour
60g millet flour, plus extra
    for flouring the surface
25g sorghum flour
10g salt
5g xanthan gum
2 teaspoons apple cider vinegar ②

Imagine a perfect loaf of bread – with a deep golden-brown crust, all crisp and caramelised, so that it makes an inviting 'crunch' as you cut into it. The cut slice reveals a soft, yet chewy interior and an open crumb, accompanied by that wonderful yeasty bread smell. You can eat it as it is, fry it in a bit of butter to golden perfection, maybe make a toasted cheese sandwich… tell me, is your mouth watering yet? Good. What you've just imagined is this artisan white crusty loaf.

+   In a small bowl, mix together the yeast, sugar and 170g warm water. Set aside for 10–15 minutes or until the mixture starts frothing.
+   In a separate bowl, mix together the psyllium husk and the remaining 200g water. After about 15–30 seconds, a gel will form.
+   In a large bowl, mix together the tapioca starch, brown rice flour, millet flour, sorghum flour, salt and xanthan gum.
+   Add the yeast mixture, psyllium gel and apple cider vinegar. By hand or using a stand mixer fitted with the dough hook, knead the dough until it is smooth and starts coming away from the bowl (about 5–10 minutes).
+   Transfer the dough to a lightly oiled surface and, with lightly oiled hands, gently flatten the dough into a disc. Then, take sections around the edge and fold them inwards over themselves, rotating the dough as you go. Once you have completed a 360° rotation, flip the dough so that the seam is underneath, leaving you with a smooth dough ball.
+   Place the dough into a lightly oiled bowl, seam side downwards, cover with a tea towel and allow to rise in a warm place for about 1 hour or until doubled in volume.
+   Lightly dust your work surface with millet flour. Once the dough has risen, turn it out on to the floured surface and use the heel of your hand to knead it into shape, folding sections of the dough back over themselves, rotating as you go. If the surface isn't smooth after the first full rotation, continue kneading until you're happy with how it looks.
+   Flip it seam side downwards on to an unfloured section of the work surface and gently rotate in place to seal the seams.
+   Dust an 18cm proving basket with millet flour and place the dough inside with the seams facing upwards. ③ Cover with a tea towel and prove in a warm place for about 1 hour until doubled in volume.
+   While the loaf is proving, place a cast-iron skillet on the middle shelf of the oven and a baking tray on the bottom shelf; or a Dutch oven or combo cooker on the lower–middle shelf (no need for the baking tray). Close the door and preheat the oven to 250°C.
+   Once the dough has doubled in volume, turn it out of the basket on to a piece of baking paper and, using a lame or a sharp knife, score the top (the easiest pattern is a criss-cross, about 0.75–1cm deep).
+   Take the hot cast-iron skillet, or Dutch oven or combo cooker, out of the oven and transfer the bread along with the baking paper into it. To do

this, for a skillet or combo cooker, slide a pizza peel or baking sheet underneath the baking paper and slide the bread along with the baking paper gently into the hot skillet or combo cooker. For a Dutch oven, use the sides of the baking paper as handles to transfer the bread into it.

+ For a cast-iron skillet: Return the cast-iron skillet to the oven, pour some boiling water into the hot tray at the bottom of the oven, place 3 ice cubes around the sides of the bread (between the baking paper and the skillet walls) and immediately close the oven door. ④ Bake with steam for 20 minutes – don't open the oven door, as that would allow the steam to escape.
+ For a Dutch oven or combo cooker: Place 3 ice cubes around the sides of the bread (between the baking paper and the container walls), and immediately close the lid on top. Return to the oven and bake, covered, for 20 minutes.
+ After the 20 minutes, reduce the oven temperature to 230°C, remove the water-filled tray from the oven (for the cast-iron skillet) or uncover the Dutch oven or combo cooker, and bake, steam-free, for a further 40 minutes. The final loaf should be deep, dark brown and weigh about or below 640g (about 16% weight loss). If the loaf browns too quickly, cover with foil (shiny side up) and bake until done.
+ Remove from the oven and transfer to a wire rack to cool completely – the final weight should be about 630g (about 17% weight loss).

STORAGE
3–4 days in a closed container in a cool, dry place. Although it's best to toast the bread on days 3 and 4.

Baker's percentage (b%), ratios and weight loss for white crusty loaf

QUANTITY NAME	VALUE
hydration	112 b%
starch	38 b%
binders	6 b%
psyllium:xanthan ratio	3:1
expected % weight loss after baking and cooling	~17%

PHOTOGRAPHY OVERLEAF ⟶

NOTES
① *Because this loaf uses both psyllium husk and xanthan gum, its crumb is softer, fluffier and not quite as chewy as that of the artisan dark crusty loaf (see page 272), which uses only psyllium husk.*

② *The apple cider vinegar helps to form the loaf's open crumb by creating a slightly acidic environment for the yeast, making it even more active. The acidity of the vinegar is, in a way, mimicking that of a sourdough starter.*

③ *If you don't have a proving basket, you can use a bowl (18cm in diameter) lined with a clean tea towel.*

④ *The two steam sources start generating steam at different times. The ice cubes provide an immediate source of steam close to the loaf, which gives the water in the tray below enough time to start evaporating. The combination of both gives the best results.*

ARTISAN DARK CRUSTY LOAF

MAKES 1 loaf
PREP TIME 45 mins
RISE TIME 2 hours
BAKE TIME 1 hour

Compared to its white counterpart (see page 266), this bread is richer in flavour, thanks to the addition of white teff and buckwheat flours, which give it a deep, almost nutty taste. The high hydration and the addition of apple cider vinegar result in a dark, crunchy crust and a chewy, open crumb. The loaf stays moist for days – although whether it will actually last that long without being eaten is another story.

8g active dried yeast
20g caster sugar
390g warm water
20g psyllium husk ①
100g tapioca starch
90g millet flour, plus extra for
    flouring the surface
90g white teff flour
50g buckwheat flour
10g salt
2 teaspoons apple cider vinegar ②

+ In a small bowl, mix together the yeast, sugar and 150g warm water. Set aside for 10–15 minutes or until the mixture starts frothing.
+ In a separate bowl, mix together the psyllium husk and the remaining 240g water. After about 15–30 seconds, a gel will form.
+ In a large bowl, mix together the tapioca starch, millet flour, white teff flour, buckwheat flour and salt.
+ Add the yeast mixture, psyllium gel and apple cider vinegar. By hand or using a stand mixer fitted with a dough hook, knead the dough until it is smooth and starts coming away from the bowl (about 5–10 minutes).
+ Transfer the dough to a lightly oiled surface and, with lightly oiled hands, shape it into a ball. To do this, gently flatten the dough into a disc. Then, take sections around the edge and fold them inwards over themselves, rotating the dough as you go. Once you have completed a 360° rotation, flip the dough so that the seam is underneath, leaving you with a smooth dough ball.
+ Place the dough into a lightly oiled bowl, seam side downwards, cover with a tea towel and allow to rise in a warm place for about 1 hour or until doubled in volume.
+ Lightly dust your work surface with millet flour. Once the dough has risen, turn it out on to the floured surface and use the heel of your hand to knead it into shape, folding sections of the dough back over themselves, rotating as you go. If the surface isn't smooth after the first full rotation, continue kneading until you're happy with how it looks.
+ Flip it seam side downwards on to an unfloured section of the work surface and gently rotate in place to seal the seams.
+ Dust an 18cm proving basket with some millet flour and place the dough inside, seams upwards. ③ Cover with a tea towel and prove in a warm place for about 1 hour until doubled in volume.
+ While the loaf is proving, place a cast-iron skillet on the middle shelf of the oven and a baking tray on the bottom shelf; or a Dutch oven or combo cooker on the lower–middle shelf (no need for the baking tray). Close the door and preheat the oven to 250°C.
+ Once the dough has doubled in volume, gently turn it out of the basket on to a piece of baking paper and use a lame or a sharp knife to score the top (the easiest pattern is a criss-cross, about 0.75–1cm deep).
+ Take the hot cast-iron skillet, or Dutch oven or combo cooker out of the oven and transfer the bread along with the baking paper into it. To do

this, for a skillet or combo cooker, slide a pizza peel or baking sheet underneath the baking paper and slide the bread along with the baking paper gently into the hot skillet or combo cooker. For a Dutch oven, use the sides of the baking paper as handles to transfer the bread into it.

+ For a cast-iron skillet: Return the cast-iron skillet to the oven, pour some boiling water into the hot tray at the bottom of the oven, place 3 ice cubes around the sides of the bread (between the baking paper and the skillet walls) and immediately close the oven door. ④ Bake with steam for 20 minutes – don't open the oven door, as that would allow the steam to escape.
+ For a Dutch oven or combo cooker: Place 3 ice cubes around the sides of the bread (between the baking paper and the cast-iron container walls), and immediately close the lid on top. Transfer into the oven and bake, covered, for 20 minutes.
+ After the 20 minutes, reduce the oven temperature to 230°C, remove the water-filled tray (for the cast-iron skillet) or uncover the Dutch oven or combo cooker, and bake, steam-free, for a further 40 minutes. The final loaf should be deep dark brown and weigh about or below 640g (about 19% weight loss). If the loaf browns too quickly, cover with foil (shiny side up) and bake until done.
+ Remove from the oven and transfer to a wire rack to cool completely – the final weight should be about 630g (about 20% weight loss).

STORAGE
3–4 days in a closed container in a cool, dry place. Although it's best to toast the bread on days 3 and 4.

Baker's percentage (b%) and weight loss for dark crusty loaf

QUANTITY NAME	VALUE
hydration	118 b%
starch	30 b%
binders	6 b%
expected % weight loss after baking and cooling	~20%

PHOTOGRAPHY OVERLEAF ⟶

NOTES

① *Because this loaf uses only psyllium husk and no xanthan gum, its crumb is chewier than that of the artisan white crusty loaf (see page 266), which uses a combination of both binders.*

② *The apple cider vinegar helps to form the open crumb of the loaf by creating a slightly acidic environment for the yeast, making it even more active. The acidity of the vinegar is, in a way, mimicking that of a sourdough starter.*

③ *If you don't have a proving basket, you can use a bowl (about 18cm in diameter) lined with a clean tea towel.*

④ *The two steam sources start generating steam at different times. The ice cubes provide an immediate source of steam close to the loaf, which gives the water in the tray below enough time to start evaporating. The combination of both gives the best results.*

SANDWICH BREAD

MAKES 1 loaf
PREP TIME 45 mins
COOK TIME 5 mins
RISE TIME 1 hour 30 mins
BAKE TIME 1 hour

The tangzhong method of making dough (see page 264) takes this buttery bread, perfect for sandwiches and toasting, all the way to pillowy soft perfection. The crust is golden brown and benefits from a generous glaze of butter – both before and after baking. It's a versatile bread, and is easily sliceable thanks to being baked in a loaf tin. Use it for everything from a grilled cheese sandwich to French toast.

GLUTEN-FREE FLOUR MIX
220g tapioca starch
185g millet flour, plus extra
    for flouring the surface
35g sorghum flour
7g xanthan gum ①

TANGZHONG
45g gluten-free flour mix
    (see above)
105g water

SANDWICH BREAD
10g active dried yeast
25g caster sugar
360g warm water
20g psyllium husk
1 quantity of tangzhong
    (see above)
remaining quantity of gluten-free
    flour mix (see above)
8g salt
40g unsalted butter, softened,
    plus extra (melted) for brushing
    the bread

GLUTEN-FREE FLOUR MIX
+ In a large bowl, thoroughly mix together all the ingredients for the gluten-free flour mix.

TANGZHONG
+ In a small saucepan, combine the portion of the gluten-free flour mix with the water and mix well to combine. Don't worry if you have a few flour clumps, though, as they will disappear during cooking.
+ Cook the mixture over a medium–high heat, stirring continuously, until it forms a thick slurry or paste (about 5 minutes).
+ Remove from the heat, transfer to a small bowl, cover with cling film, and set aside to cool until needed.

SANDWICH BREAD
+ In a small bowl, mix together the yeast, sugar and 140g warm water. Set aside for 10–15 minutes or until the mixture starts frothing.
+ In a separate bowl, mix together the psyllium husk and the remaining 220g water. After about 15–30 seconds, a gel will form.
+ In a large bowl, combine the yeast mixture, psyllium gel, tangzhong, the remaining gluten-free flour mix and the salt. Using a stand mixer fitted with the dough hook or by hand, knead the dough until smooth (about 10–15 minutes). The dough will be quite sticky and will stick to the bottom and sides of the bowl.
+ Add the softened butter and knead for a further 5 minutes until all the butter has been fully incorporated.
+ Transfer the dough to a lightly oiled container. Allow to rise in a warm place for about 1 hour or until doubled in volume. Meanwhile, you'll need a 900g loaf tin (23 x 13 x 7.5cm) for the next steps.
+ Turn out the risen dough on to a lightly floured surface and roll it out into a 20 x 30cm rectangle. Take a short end and roll up the dough along its length, to give you a 20cm loaf. Transfer the loaf into the loaf tin and prove for about 30–45 minutes or until doubled in volume.
+ While the sandwich bread is proving, preheat the oven to 200°C.

+ Brush the risen sandwich bread with melted butter and bake for about 1 hour or until golden brown. If it starts browning too quickly, cover with foil (shiny side up) and bake until done. The loaf should weigh about 890g (about 12% weight loss).

+ As soon as the loaf comes out of the oven, brush it with some more melted butter. ② Transfer to a wire rack and allow to cool completely – the final weight after cooling will be about 880g (about 13% weight loss).

STORAGE

3–4 days in a closed container in a cool dry, place. Although it's best to toast the bread on days 3 and 4.

Baker's percentage (b%), ratios and weight loss for sandwich bread

QUANTITY NAME	VALUE
hydration	106 b%
starch	50 b%
binders	6 b%
psyllium:xanthan ratio	~3:1
expected % weight loss after baking and cooling	~13%

NOTES

① *The xanthan gum gives the sandwich bread a softness and fluffiness that would be impossible to achieve with psyllium husk alone.*

② *The butter glaze makes the crust much softer, and gives it a richer flavour.*

PHOTOGRAPHY OVERLEAF ⟶

THE PERFECT BAGUETTES

MAKES 2 baguettes (32cm each)
PREP TIME 45 mins
RISE TIME 1 hour 20 mins
BAKE TIME 40 mins

8g active dried yeast
20g caster sugar
390g warm water
20g psyllium husk
130g tapioca starch
100g millet flour, plus extra
    for flouring the surface
50g white teff flour
50g buckwheat flour
10g salt
2 teaspoons apple cider vinegar ①
2 tablespoons melted butter, for
    glazing ②

What makes the perfect baguette? An impressive, deep golden-brown, crispy crust; a soft, chewy interior filled with holes; an inviting smell that makes you itch to tear off a piece, smear a generous pat of butter on top and take that first bite. Well, these baguettes tick all of those boxes (I can attest to that last one), and do so effortlessly – thanks to the high hydration of the dough, the addition of apple cider vinegar and the butter glaze, which keeps the crust nice and golden. Introducing steam into the oven by spraying the baguettes with water in the first 10 minutes of baking keeps the crust malleable enough for maximum oven spring to occur.

+ In a small bowl, mix together the yeast, sugar and 150g warm water. Set aside for 10–15 minutes or until the mixture starts frothing.
+ In a separate bowl, mix together the psyllium husk and 240g water. After about 15–30 seconds, a gel will form.
+ In a large bowl, mix together the tapioca starch, millet flour, white teff flour, buckwheat flour and salt.
+ Add the yeast mixture, psyllium gel and apple cider vinegar. Using a stand mixer fitted with a dough hook or by hand, knead the dough until it is smooth and starts coming away from the bowl (about 5–10 minutes).
+ Transfer the dough to a lightly oiled surface and, with lightly oiled hands, gently flatten the dough into a disc. Then, take sections around the edge and fold them inwards over themselves, rotating the dough as you go. Once you have completed a 360° rotation, flip the dough so that the seam is underneath, leaving you with a smooth dough ball.
+ Place the dough into a lightly oiled bowl, seam side downwards, cover with a tea towel and allow to rise in a warm place for about 1 hour or until doubled in volume.
+ Lightly flour your work surface with millet flour. Turn out the risen dough and give it a thorough knead to deflate it. Then, divide it into 2 equal pieces.
+ Lightly flatten each piece and roll it up tightly into a log, pinching together the seams. Gently roll each log until about 30cm long. Transfer the logs, seam sides downwards, on to a baguette pan, cover lightly with a tea towel and prove in a warm place for about 20–30 minutes or until slightly risen but not quite doubled in volume. ③
+ While the baguettes are proving, preheat the oven to 230°C with a large baking sheet or baking tray on the middle shelf. Make sure the baguette pan will fit on top of the baking sheet or tray.

+ Brush the proved baguettes gently with melted butter to glaze and, using a lame or a sharp knife, cut 3–5 diagonal slashes on top of each baguette. Bake for 10 minutes, spraying the baguettes with water three times during this initial period, making sure that you open the oven door as little as possible. ④
+ After the 10 minutes, reduce the oven temperature to 200°C and bake for a further 30–35 minutes. If the baguettes brown too quickly, cover with foil (shiny side up) and bake until done. The final baguettes should be deep golden-brown and weigh about or below 620g (about 21% weight loss).
+ Allow the baguettes to cool in the baguette pan for 10–15 minutes, then transfer them to a wire rack to cool completely – the final weight after cooling will be about 610g (about 23% weight loss).

STORAGE

Best on the day, but you can keep them in an airtight container until the following day if necessary.

Baker's percentage (b%) and weight loss for perfect baguettes

QUANTITY NAME	VALUE
hydration	118 b%
starch	39 b%
binders	6 b%
expected % weight loss after baking and cooling	~23%

NOTES

① *The apple cider vinegar helps to form the open crumb of the baguettes by creating a slightly acidic environment for the yeast, making it even more active. The acidity of the vinegar is, in effect, mimicking that of a sourdough starter.*

② *Without the butter glaze, the baguettes can sometimes form a white-ish, dull and dry crust. The butter glaze makes that crust beautifully golden. You can use sunflower or olive oil, if you prefer.*

③ *If you don't have a baguette pan, place the unproved baguettes on to separate pieces of baking paper and then on to a tea towel gathered to form folds on either side of each baguette (like a professional baguette couche). Your proved baguettes won't have quite the same rounded shape as they would with a baguette pan, but the results will be more than good enough. Once proved, glaze them with butter, slash them and transfer them directly on to the hot baking sheet along with the baking paper.*

④ *Directly spraying the baguettes with water keeps their crusts softer and more malleable, allowing for maximum oven spring – which helps to create an open crumb with plenty of holes.*

PHOTOGRAPHY OVERLEAF ⟶

SEEDED BUNS

10g active dried yeast
20g caster sugar
380g warm water
20g psyllium husk
150g tapioca starch
120g millet flour, plus extra
    for flouring the surface
60g brown rice flour
8g salt
80g mixed seeds, such as
    pumpkin, sunflower, flax
    and poppy seeds

These simple buns are great for making sandwiches or as an accompaniment to lunch or dinner, but I honestly like them best just as they are, still slightly warm (or toasted), with butter on top.

+ In a small bowl, mix together the yeast, sugar and 140g warm water. Set aside for 10–15 minutes or until the mixture starts frothing.
+ In a separate bowl, mix together the psyllium husk and remaining 240g water. After about 15–30 seconds, a gel will form.
+ In a large bowl, mix together the tapioca starch, millet flour, brown rice flour and salt. Add the yeast mixture and the psyllium gel.
+ Using a stand mixer fitted with the dough hook or by hand, knead the dough until it is smooth and starts coming away from the bowl (about 5–10 minutes). Reserve about 2 tablespoons of the seeds and add the remainder to the dough. Mix for 2–3 minutes until fully incorporated.
+ Allow the dough to rise in a warm place for about 1 hour or until doubled in volume. Turn out the risen dough on to a lightly floured surface and gently pat it down into a rough 15 x 20cm rectangle, 2.5–3.5cm thick. Divide into 6 equal buns, and transfer them to a lined baking sheet, spacing them at least 2.5cm apart. Prove the buns in a warm place for 30–45 minutes or until about doubled in volume.
+ While the buns are proving, adjust the oven shelf to the middle position and preheat the oven to 230°C.
+ Lightly spray the risen buns with water and sprinkle with the reserved seeds. Bake for about 30 minutes or until golden brown and the bases sound hollow when tapped. The combined weight of the 6 buns immediately out of the oven should be about 680g (20% weight loss).
+ Transfer the buns to a wire rack to cool – the final combined weight after complete cooling will be about 670g (about 21% weight loss).

STORAGE
3–4 days in an airtight container. Best toasted on days 3 and 4.

Baker's percentage (b%) and weight loss for seeded buns

QUANTITY NAME	VALUE
hydration	115 b%
starch	45 b%
binders	6 b%
expected % weight loss after baking and cooling	~21%

BURGER BUNS

MAKES 6
PREP TIME 45 mins
COOK TIME 5 mins
RISE TIME 1 hour 45 mins
BAKE TIME 25 mins

The magic of a good burger bun lies in a shiny crust, the interior that is a delicious contradiction of rich and buttery yet at the same time light and fluffy, and its ability to be toasted to golden, caramelised perfection. These burger buns tick all the boxes, and do so effortlessly. The tangzhong method of making the dough (see page 264) plays a central role in this recipe, guaranteeing a dough that is a joy to handle and burger buns that stay soft for days.

GLUTEN-FREE FLOUR MIX

165g tapioca starch
140g millet flour, plus extra
  for flouring the surface
25g sorghum flour
5g xanthan gum ①

TANGZHONG

35g gluten-free flour mix
  (see above)
80g water

BURGER BUN DOUGH

10g active dried yeast
20g caster sugar
200g warm water
15g psyllium husk
1 egg, at room temperature
1 quantity of tangzhong
  (see above)
remaining quantity of gluten-free
  flour mix (see above)
6g salt
30g unsalted butter, softened

TO ASSEMBLE

1 egg, at room temperature,
  whisked
1 tablespoon sesame seeds
2 tablespoons melted butter

GLUTEN-FREE FLOUR MIX

+ In a large bowl, thoroughly mix together all the ingredients for the gluten-free flour mix.

TANGZHONG

+ In a small saucepan, combine the portion of the gluten-free flour mix with the water and mix well to combine. Don't worry if you have a few flour clumps, though, as they will disappear during cooking.
+ Cook the mixture over a medium–high heat, stirring continuously, until it forms a thick slurry or paste (about 5 minutes).
+ Remove from the heat, transfer to a small bowl, cover with cling film, and set aside to cool until needed.

BURGER BUN DOUGH

+ In a small bowl, mix together the yeast, sugar and 80g warm water. Set aside for 10–15 minutes or until the mixture starts frothing.
+ In a separate bowl, mix together the psyllium husk and the remaining 120g water. After about 15–30 seconds, a gel will form.
+ In a large bowl, mix together the yeast mixture, psyllium gel, egg, tangzhong, remaining gluten-free flour mix and salt. Using a stand mixer fitted with the dough hook or by hand, knead the dough until smooth (about 5–10 minutes). It will be quite sticky and will stick to the bottom and sides of the bowl.
+ Add the softened butter and knead for a further 5 minutes until all the butter has been fully incorporated.
+ Transfer the dough to a lightly oiled container and allow to rise in a warm place for 1 hour–1 hour 30 minutes or until doubled in volume.
+ Turn out the risen dough on to a lightly floured surface and divide it into 6 equal portions. Shape each portion into a ball, and place on a baking sheet lined with baking paper, at least 5cm apart. Prove in a warm place for about 45 minutes or until doubled in volume.

+ While the burger buns are proving, adjust the oven shelf to the middle position and preheat the oven to 200°C.
+ Once the burger buns are proved, brush them with whisked egg, then sprinkle with the sesame seeds. Bake for about 25–30 minutes or until golden brown. The combined weight of the buns immediately out of the oven should be about 650g (about 13% weight loss).
+ Remove the buns from the oven and immediately brush them with the melted butter. ② Transfer to a wire rack to cool before using. The final weight after cooling will be about 640g (about 14% weight loss).

STORAGE
3–4 days in an airtight container in a cool, dry place. Although it's best to toast the buns on days 3 and 4.

Baker's percentage (b%), ratios and weight loss for burger buns

QUANTITY NAME	VALUE
hydration	85 b%
starch	50 b%
binders	6 b%
psyllium:xanthan ratio	3:1
expected % weight loss after baking and cooling	~14%

NOTES
① *The xanthan gum gives the burger buns a softness and fluffiness that would be impossible to achieve with psyllium husk alone.*

② *The butter glaze makes the crusts much softer, and gives the buns an additional, glossy shine.*

PHOTOGRAPHY OVERLEAF ⟶

MEDITERRANEAN PULL-APART ROLLS

SERVES 8
PREP TIME 45 mins
RISE TIME 1 hour 45 mins
BAKE TIME 45 mins

10g active dried yeast
20g caster sugar
340g warm water
15g psyllium husk
125g tapioca starch
120g brown rice flour
60g millet flour, plus extra for
    flouring the surface
25g sorghum flour
6g salt
5g xanthan gum
50g pitted green or black olives,
    sliced
30g sun-dried tomatoes in oil,
    drained and chopped
polenta, for sprinkling
olive oil, for glazing

I'm not quite sure what it is about adding olives to bread, but doing so makes the flavour about a hundred times better. These fluffy, soft Mediterranean pull-apart rolls, with olives, sun-dried tomatoes and a glaze of olive oil, are a perfect example.

+ In a small bowl, mix together the yeast, sugar and 150g warm water. Set aside for 10–15 minutes or until the mixture starts frothing.
+ In a separate bowl, mix together the psyllium husk and the remaining 190g water. After about 15–30 seconds, a gel will form.
+ In a large bowl, mix the tapioca starch, brown rice flour, millet flour, sorghum flour, salt and xanthan gum. Add the yeast mixture and the psyllium gel. Using a stand mixer fitted with the dough hook or by hand, knead the dough until smooth and it starts coming away from the bowl (about 5–10 minutes). Incorporate the olives and tomatoes.
+ Allow the dough to rise in a warm place for about 1 hour until doubled in volume. Turn out the dough on a lightly floured surface and divide it into 8 equal pieces. Shape each piece into a ball. Lightly sprinkle a 20cm round baking tin with polenta and fit the balls snugly inside. Prove in a warm place for 45–60 minutes or until about doubled in volume.
+ While the buns are proving, adjust the oven shelf to the middle position and preheat the oven to 200°C.
+ Brush the tops of the buns with olive oil, sprinkle with polenta, and bake for 45–50 minutes or until golden brown. The finished rolls should have a combined weight of about 700g (13% weight loss). If the rolls brown too quickly, cover with foil (shiny side up) and bake until done.
+ Remove from the oven and allow to cool in the tin for 5–10 minutes, then transfer to a wire rack to cool completely. The final weight after cooling will be about 690g (about 14% weight loss).

STORAGE
3–4 days in a closed container in a cool, dry place.

Baker's percentage (b%), ratios and weight loss for pull-apart rolls

QUANTITY NAME	VALUE
hydration	103 b%
starch	38 b%
binders	6 b%
psyllium:xanthan ratio	3:1
expected % weight loss after baking and cooling	~14%

PROPER BOILED + BAKED BAGELS

MAKES 8
PREP TIME 1 hour
RISE TIME 1 hour 30 mins
COOK TIME 2 x 2 mins
BAKE TIME 25 mins

The days of missing bagels are over. These gluten-free beauties are the real deal – with a crispy, golden crust and a chewy (yet somehow soft and pillowy) interior. What's more, they're not just bread baked with a hole in the middle. Oh no – these are boiled and baked bagels, and every bite will transport you into the best bagel shops of NYC.

8g active dried yeast
20g caster sugar
160g whole milk, warmed
12g psyllium husk
160g water, at room temperature
140g tapioca starch
80g brown rice flour
80g millet flour, plus extra for
  flouring the surface
8g xanthan gum
8g salt
1 tablespoon molasses ①
1 egg, whisked
1 tablespoon poppy seeds
1 teaspoon flaky sea salt
fillings of choice, to serve

+ In a small bowl, mix together the yeast, sugar and warm milk. Set aside for 10–15 minutes or until the mixture starts frothing.
+ In a separate bowl, mix together the psyllium husk and water. After about 15–30 seconds, a gel will form.
+ Whisk together the tapioca starch, brown rice flour, millet flour, xanthan gum and salt. Add the yeast mixture and the psyllium gel.
+ Using a stand mixer fitted with the dough hook or by hand, knead the dough until it's smooth and starts coming away from the bowl (about 5–10 minutes). Allow to rise in a warm place for about 1 hour or until doubled in volume.
+ Divide the risen dough into 8 portions. On a lightly floured surface, shape each portion into a ball and, using your index finger, create a hole in the middle. Twirl each bagel around your finger until the hole is about 2.5cm in diameter. ② Place the bagels on a lightly oiled tray and prove in a warm place for 30 minutes or until risen by about half.
+ Bring a large pan of water to the boil, adjust the oven shelf to the middle position, preheat the oven to 200°C and line a baking sheet with baking paper. Add the molasses to the pan of water. Drop 4 bagels into the water, cook them for 1 minute on each side, then remove and transfer to the lined baking sheet. ③ Repeat for the remaining 4 bagels.
+ Brush the bagels with the whisked egg, sprinkle with the poppy seeds and sea salt, and bake for about 25–30 minutes or until golden brown.
+ Transfer to a wire rack to cool. Serve sliced and filled, as you wish.

NOTES

① *Traditionally, bagels are made with barley malt syrup or extract, which contains gluten. Molasses is a wonderful alternative, and it improves the flavour and the caramelisation of the bagel crust.*

② *Because the psyllium husk and xanthan gum don't completely recreate the strong gluten network of a regular bagel, I recommend you twirl the bagels on the floured surface rather than in the air.*

③ *Boiling sets the crust and halts the rise of the bagels, giving them the typically chewy interior.*

STORAGE
Best on the day; or, store in an airtight container and toast the next day.

Baker's percentage (b%) and ratios for bagels

QUANTITY NAME	VALUE
hydration	107 b%
starch	47 b%
binders	7 b%
psyllium:xanthan ratio	3:2

THE PERFECT PIZZA

MAKES 2 (each 23cm diameter)
PREP TIME 1 hour
RISE TIME 1 hour 45 mins
BAKE TIME 2 x 18 mins

PIZZA DOUGH

7g active dried yeast
15g caster sugar
340g warm water
18g psyllium husk
140g tapioca starch
120g millet flour, plus extra for
    flouring
40g sorghum flour
8g salt
3 tablespoons olive oil
2 teaspoons apple cider vinegar ①

TOMATO SAUCE
+ TOPPINGS

200g tinned plum tomatoes
½ teaspoon salt
¼ teaspoon pepper
1–2 teaspoons dried oregano
100–150g grated cheddar, sliced
    mozzarella, or other cheese
    of choice
8–12 basil leaves, to serve

This gluten-free pizza recipe is my pride and joy. It has the authentic, crisp bottom and an open crumb with plenty of holes in the edge crust. It's a real-deal, mouthwatering, finger-licking delicious pizza. It also handles very similarly to its wheat-based counterpart: you can shape and stretch it like you would wheat dough – with the caveat that you do so a bit more carefully in the absence of gluten. Still, the psyllium husk provides plenty of elasticity, and even if a tear happens here or there, you can easily patch it up (or leave it and call it rustic charm). My very favourite pizzas are those that are a bit wonky, with a few dark, almost charred patches on the crust – those that make you feel like you're out in a quaint little trattoria somewhere in Italy, eating a pizza straight out of a wood-fired oven.

PIZZA DOUGH

+ In a small bowl, mix together the yeast, sugar and 160g warm water. Set aside for 10–15 minutes or until the mixture starts frothing.
+ In a separate bowl, mix together the psyllium husk and the remaining 180g water. After about 15–30 seconds, a gel will form.
+ In a large bowl, mix together the tapioca starch, millet flour, sorghum flour and salt. Add the yeast mixture, psyllium gel, 1 tablespoon of the olive oil and the apple cider vinegar. Using a stand mixer fitted with the dough hook or by hand, knead the dough until it's smooth and starts coming away from the bowl (about 5–10 minutes).
+ Transfer the bread to a lightly oiled surface and knead it gently, forming it into a smooth ball. Allow the dough to rise in a warm place for about 1 hour or until doubled in volume.
+ Divide the risen dough into 2 equal pieces and shape each half into a tight ball. Don't worry if the surface of the dough isn't completely smooth. Place the dough balls on a lightly millet-floured tray, cover with a tea towel and prove in a warm place for 30–45 minutes.

TOMATO SAUCE

+ Crush the tinned plum tomatoes into a rough purée, using either your hands or a food blender.
+ Mix in the salt, pepper and oregano, then set aside until needed.

SHAPING + ASSEMBLING THE PIZZA

+ Place a large baking tray or sheet on the upper-middle shelf of the oven and preheat the oven to 250°C. ②
+ Generously flour your work surface with millet flour. Place a proved pizza dough ball on the work surface and press down with your fingers to create a well in the middle of the ball. Use your knuckles to gently stretch the dough into a circle, rotating the dough as you go. You can also carefully lift the dough up into the air, holding it with both hands

on one edge and letting the rest of the dough hang down. This way, gravity will take care of the stretching for you as you rotate the dough like you would turn a wheel, holding it by the edge all the time.

+ Keep turning and stretching until the dough gives you a rough 23cm-diameter circle. It should be about 2mm thick in the middle with a 1–2.5cm border that's about 1–1.5cm thick. The dough might tear slightly at the edges or have holes in the middle – the edge tears will close up in the oven as the crust puffs up and you can close up the holes in the middle by pinching together the surrounding dough.
+ Allow the pizza base to prove on the work surface for 15–20 minutes. ③
+ Place the rested pizza base on a floured pizza peel or baking sheet. Brush the edges with half the remaining olive oil and spoon half the tomato sauce on top, spreading it evenly across the pizza up to the border. ④
+ Slide the pizza from the peel or baking sheet on to the hot baking tray or sheet and bake for 15–18 minutes until the edges are a dark golden brown. While the first pizza is baking, prepare the second.
+ Take the pizza out of the oven, scatter the grated cheese on top, and return to the oven to bake for a further 3–5 minutes until the cheese is bubbling. ⑤ Repeat the baking steps for the second pizza.
+ Allow to cool for about 5 minutes, then top with fresh basil leaves, slice and serve.

STORAGE
Eat as soon as possible after baking, while still hot or warm.

Baker's percentage (b%) for pizza dough

QUANTITY NAME	VALUE
hydration	113 b%
starch	47 b%
binders	6 b%

NOTES

① *The apple cider vinegar creates a slightly acidic environment for the yeast, boosting its activity and in turn helping to form the characteristic open crumb of the pizza crust. In a way, the acidity mimics that of a sourdough starter.*

② *The high temperature and a hot baking tray or sheet are crucial for a crisp pizza base. If you have a baking stone, use that instead.*

③ *This short, final proving before assembly and baking opens up the crumb of the outer pizza crust even further, resulting in plenty of holes and a soft texture.*

④ *The oil glaze is mostly for aesthetics – without it, the gluten-free pizza crust sometimes has a slightly dull, white-ish appearance. A quick brush with some olive oil guarantees golden-brown-crust perfection.*

⑤ *Baking the pizza with only the tomato sauce and then adding the cheese for the last few minutes guarantees a crisp base and perfectly melted, bubbly cheese. Add other toppings such as olives or pepperoni at the same time as the tomato sauce, if you wish.*

PHOTOGRAPHY OVERLEAF ⟶

ROSEMARY FOCACCIA

SERVES 6–8
PREP TIME 45 mins
RISE TIME 1 hour 30 mins
BAKE TIME 50 mins

A perfect focaccia should have a crisp, oil-soaked, golden crust; a soft, open crumb; and a deliciously chewy texture. In this recipe, psyllium husk gives elasticity and structure; olive oil in the dough keeps the crumb chewy and the flavour rich; and high hydration and the apple cider vinegar give an open crumb with plenty of holes.

3 rosemary sprigs, each about
  10cm long
4 tablespoons olive oil
8g active dried yeast
20g caster sugar
340g warm water
18g psyllium husk
140g tapioca starch
120g millet flour, plus extra for
  flouring the surface
40g sorghum flour
6g salt
2 teaspoons apple cider vinegar ①
flaky sea salt, for sprinkling

NOTES

① *The apple cider vinegar creates a slightly acidic environment for the yeast, making it even more active and in turn helping to form the characteristic open crumb of the focaccia. In a way, the acidity mimics that of a sourdough starter.*

② *This focaccia takes significantly longer than a gluten-based focaccia to cook through. Don't be tempted to take it out of the oven before the 50 minutes are up (no matter how golden brown it looks), as you'll be left with an underbaked, dense interior.*

+ Tear or cut the rosemary sprigs into 1cm pieces and mix them with 2 tablespoons of the olive oil. Set aside until needed.
+ In a small bowl, mix together the yeast, sugar and 160g warm water. Set aside for 10–15 minutes or until the mixture starts frothing.
+ In a separate bowl, mix together the psyllium husk and the remaining 180g water. After about 15–30 seconds, a gel will form.
+ In a large bowl, mix together the tapioca starch, millet flour, sorghum flour and salt. Add the yeast mixture, psyllium gel, the remaining olive oil and the apple cider vinegar. Using a stand mixer fitted with the dough hook or by hand, knead the dough until it is smooth and starts coming away from the bowl (about 5–10 minutes). Allow the dough to rise in a warm place for about 1 hour or until doubled in volume.
+ Line a large baking sheet or baking tray with baking paper.
+ Turn out the dough on to a lightly floured surface and roll it out to a 20 x 28cm rectangle, about 1cm thick. Transfer to the baking tray or sheet, gently stretching it out if necessary. Cover with a tea towel and prove in a warm place for about 30–45 minutes until puffed up. Adjust the oven shelf to the middle position and preheat the oven to 220°C.
+ Use your fingers to dimple the risen focaccia. Push the oil-soaked rosemary pieces into the dimples, drizzle with the rosemary-infused olive oil and sprinkle with salt. Bake for 50–55 minutes until golden brown. ② If the top browns too quickly, cover with foil (shiny side up) and bake until done. Remove from the oven, cool on the baking sheet for 5–10 minutes, then transfer to a wire rack to cool until warm.

STORAGE

Best warm or on the day of baking. Or, keep in an airtight container until the following day and reheat in the microwave for 10–15 seconds to serve.

Baker's percentage (b%) for focaccia

QUANTITY NAME	VALUE
hydration	113 b%
starch	47 b%
binders	6 b%

QUICK + EASY FLAT BREADS

MAKES 5
PREP TIME 15 mins
COOK TIME 10 mins

16g psyllium husk
160g whole milk
100g tapioca starch
80g brown rice flour
40g millet flour, plus extra for
   flouring the surface
1 teaspoon baking powder
¼ teaspoon salt
50g full-fat plain yoghurt
20g olive oil, plus extra for frying
   and brushing

NOTE

① *Psyllium husk doesn't form as
strong a gel with milk as it does
with water – it both takes longer
to form and is less firm in texture.
However, it still provides the flat
bread with enough flexibility
and elasticity to be workable.*

On the table in less than half an hour, these flat breads make the
perfect addition to a quick lunch or dinner. They are wonderfully
soft and flexible, and they don't crack or tear even when packed to
the brim with various fillings, making them perfect for wraps.

+ In a small bowl, mix together the psyllium husk and milk. After about
  2–3 minutes, a gel will form. ①
+ In a larger bowl, mix together the tapioca starch, brown rice flour,
  millet flour, baking powder and salt.
+ Add the psyllium gel, yoghurt and olive oil, and knead everything
  together to a smooth dough.
+ Divide the dough into 5 equal pieces and shape them into balls.
  On a surface lightly dusted with millet flour, roll out the balls into
  rough circles about 18–20cm in diameter and about 1–2mm thick.
+ Heat up a frying pan over a high heat, brush it lightly with olive oil
  and cook the flat breads one at a time. Cook the first flat bread for about
  1 minute or until slightly puffed up with brown spots on the underside.
  Brush the top lightly with olive oil, flip, and cook on the other side for
  a further 1 minute.
+ Once cooked, wrap the flat bread in a tea towel to trap in any steam
  and keep it soft and flexible. Repeat the cooking and wrapping with
  the remaining flat breads, then serve.

STORAGE

Serve warm as soon as possible after cooking.

PITTA BREADS

MAKES 6
PREP TIME 30 mins
RISE TIME 1 hour
BAKE TIME 3 x 5 mins

8g active dried yeast
10g caster sugar
230g warm water
16g psyllium husk
140g tapioca starch
40g brown rice flour
40g millet flour, plus extra for
    flouring the surface
6g salt
1 tablespoon olive oil

NOTE

① *Baking the pitta breads in the oven provides high heat on all sides to guarantee perfectly formed pockets – which you wouldn't reliably get if you cooked the pitta breads in a frying pan on the hob.*

There are few things as joyful as peering into the oven to see a pitta bread puff up to glorious heights. Not only is it a dramatic transformation from what looks like a thin pancake, it's also a visual guarantee that your pitta bread will have a generous pocket to fill with whatever deliciousness you fancy.

+   In a small bowl, mix together the yeast, sugar and 70g warm water. Set aside for 10–15 minutes or until the mixture starts frothing.
+   In a separate bowl, mix together the psyllium husk and the remaining 160g water. After about 15–30 seconds, a gel will form.
+   In a large bowl, mix together the tapioca starch, brown rice flour, millet flour and salt. Add the yeast mixture, psyllium gel and olive oil. Using a stand mixer fitted with the dough hook or by hand, knead the dough until smooth and it starts coming away from the bowl (about 5–10 minutes). The final dough will be fairly firm and stiff.
+   Transfer the dough to a lightly oiled surface and knead it gently, forming it into a smooth ball. Allow the dough to rise in a warm place for about 1 hour or until doubled in volume.
+   Place a large baking sheet on the upper-middle shelf of the oven and preheat the oven to 240°C. ①
+   Divide the risen dough into 6 equal portions, shape into balls and roll each portion out on a lightly floured surface into a rough circle about 15cm in diameter and 2–3mm thick.
+   Place the rolled-out pitta breads, 2 at a time, on the hot baking sheet and bake for 5–7 minutes or until puffed up and a light golden brown. They should start puffing up after about 3 minutes in the oven.
+   Wrap the baked pittas in a tea towel to trap in any steam and keep them soft and flexible before serving.

STORAGE
Eat warm, as soon as possible after baking.

Baker's percentage (b%) for pitta breads

QUANTITY NAME	VALUE
hydration	105 b%
starch	64 b%
binders	7 b%

CINNAMON ROLLS

SERVES 12
PREP TIME 1 hour
RISE TIME 1 hour 45 mins
BAKE TIME 35 mins

The best way to describe these cinnamon rolls is like a buttery, cinnamon-infused cloud of deliciousness, drenched in the most luxurious, velvety, cream-cheese icing. A 'buttery cloud' may sound a bit nonsensical, but it perfectly encapsulates the texture of the rolls: heavenly soft, but also incredibly moist and rich. You can make them the night before you intend to bake, if you like, making them perfect for the most amazingly lazy Sunday breakfast or brunch.

BRIOCHE DOUGH

15g active dried yeast
90g caster sugar
180g whole milk, warmed
20g psyllium husk
280g warm water
310g tapioca starch
260g millet flour, plus extra for
   flouring
50g sorghum flour
10g xanthan gum
10g salt
2 eggs, at room temperature
75g unsalted butter, softened,
   plus about 2 tablespoons
   melted butter for glazing

CINNAMON FILLING

100g unsalted butter, melted
   and cooled, ① plus about
   1 tablespoon for greasing
125g caster sugar
2–3 tablespoons ground cinnamon

CREAM-CHEESE ICING

75g full-fat cream cheese, at room
   temperature
45g unsalted butter, melted
300g icing sugar
1 teaspoon vanilla bean paste
2–3 tablespoons whole milk
   (optional)

BRIOCHE DOUGH

+ In a small bowl, mix together the yeast, 20g of the sugar and the warm milk. Set aside for 10–15 minutes or until the mixture starts frothing.
+ In a separate bowl, mix together the psyllium husk and water. After about 15–30 seconds, a gel will form.
+ In a large bowl, mix together the tapioca starch, millet flour, sorghum flour, the remaining 70g sugar and the xanthan gum and salt.
+ Add the yeast mixture, psyllium gel and eggs. Using a stand mixer fitted with the dough hook or by hand, knead everything together until you get a soft and sticky dough (about 5–10 minutes).
+ Add the softened butter and knead for a further 5 minutes until you have fully incorporated all the butter and have a very soft and sticky dough.
+ Transfer the dough into an oiled container and allow to rise in a warm place for about 1 hour–1 hour 30 minutes or until doubled in volume.

CINNAMON FILLING + ASSEMBLING THE ROLLS

+ Lightly grease a 23 x 33cm rectangular baking tin with butter.
+ In a small bowl, mix together the caster sugar and cinnamon.
+ Prepare a large sheet of baking paper and dust it generously with millet flour.
+ Turn out the proved dough on to the floured baking paper, dust it generously with millet flour, and roll it out into a 35 x 55cm rectangle. (If the dough feels too soft to handle, refrigerate it for 15–30 minutes before rolling.)
+ Brush the rolled-out dough with butter and generously sprinkle with the cinnamon–sugar mixture.
+ Roll up the dough lengthways to get a 55cm log (use the baking paper to help with the rolling) and, using a long piece of floss or a thin thread (rather than a knife, say), cut it into 12 equal portions by pressing the floss or thread all the way down to the work surface. (If the dough feels too soft to cut without drastically destroying the swirl, you can divide it into 2 logs and refrigerate them for 15–30 minutes before cutting.)

+ Place the rolls into the buttered baking tin so that they are only just touching (this gives them enough space to expand as they prove and bake). You can gently press down on them to flatten and spread them out slightly, if the pieces are a bit too tall. ②
+ Cover the baking tin with cling film (to prevent the surface from drying out) and prove the rolls in a warm place for about 45–60 minutes or until approximately doubled in volume.
+ Adjust the oven shelf to the middle position and preheat the oven to 180°C.
+ Once the rolls have proved, gently brush the tops with melted butter to glaze, then bake for about 35–40 minutes or until the tops are a light golden brown and an inserted toothpick comes out with no unbaked dough attached. If the tops brown too quickly, cover with foil (shiny side up) and bake until done.
+ For extra-rich cinnamon rolls, brush with additional melted butter immediately out of the oven, if you wish.
+ Allow to cool for 15 minutes while you make the cream-cheese icing.

CREAM-CHEESE ICING
+ Whip the cream cheese and melted butter together until smooth.
+ Add the icing sugar and vanilla paste, and whisk until smooth.
+ Depending on the desired consistency, add 2–3 tablespoons of milk and whisk well until fully combined.
+ Spread the icing evenly across the warm cinnamon rolls.

STORAGE
Best warm or within a few hours of baking. Store any leftover rolls in an airtight container for up to 1 day, then reheat in the microwave for 20–30 seconds to serve.

Baker's percentage (b%) and ratios for cinnamon rolls

QUANTITY NAME	VALUE
hydration	74 b%
starch	50 b%
binders	5 b%
psyllium:xanthan ratio	2:1

NOTES
① *The best butter consistency is very soft and spreadable – softer than the usual 'room-temperature' softened butter, but thicker than melted butter.*

② *To make **overnight cinnamon rolls**, cover the baking tin with cling film and refrigerate overnight. The following morning, prove the rolls in a warm place for about 1 hour, then bake them as instructed in the method.*

CHOCOLATE BABKA

SERVES 10–12
PREP TIME 1 hour
RISE TIME 1 hour 45 mins
CHILL TIME 30 mins
COOK TIME 5 mins
BAKE TIME 1 hour 10 mins

BRIOCHE DOUGH

8g active dried yeast
50g caster sugar
90g whole milk, warmed
10g psyllium husk
140g warm water
155g tapioca starch
130g millet flour, plus extra
   for flouring
25g sorghum flour
5g xanthan gum
5g salt
1 egg, at room temperature
35g unsalted butter, softened

CHOCOLATE FUDGE FILLING

75g dark chocolate (60–70% cocoa
   solids), chopped
45g icing sugar
40g whole milk
15g Dutch-processed cocoa
   powder
10g unsalted butter ①
¼ teaspoon salt

EGG WASH

1 egg, lightly whisked
1 tablespoon milk

SUGAR SYRUP GLAZE

25g caster sugar
1 tablespoon water

Soft, buttery dough swirled through with a luscious chocolate fudge filling – in the world of yeasted desserts, chocolate babka (a sweet braided bread of Jewish origin) has to be one of the best. To make the rolling and twisting even easier (and the end result even more beautiful), refrigerate the gluten-free brioche dough just before rolling. This sets the butter and makes the dough much easier to handle.

BRIOCHE DOUGH

+ In a small bowl, mix together the yeast, 10g of the sugar and the warm milk. Set aside for 10–15 minutes or until the mixture starts frothing.
+ In a separate bowl, mix together the psyllium husk and water. After about 15–30 seconds, a gel will form.
+ In a large bowl, mix together the tapioca starch, millet flour, sorghum flour, the remaining 40g sugar, the xanthan gum and the salt.
+ Add the yeast mixture, psyllium gel and egg. Using a stand mixer fitted with the dough hook or by hand, knead everything together to a soft and sticky dough (about 5–10 minutes).
+ Add the softened butter and knead for a further 5 minutes until you have fully incorporated all the butter. The final dough will be very soft and sticky.
+ Transfer the dough to an oiled container and allow to rise in a warm place for about 1 hour–1 hour 30 minutes or until doubled in volume.
+ Once the dough has risen, refrigerate it for about 30 minutes.

CHOCOLATE FUDGE FILLING

+ Combine all the chocolate fudge filling ingredients in a saucepan and cook over a medium–high heat until the chocolate and butter have melted and the mixture is smooth and glossy (about 5 minutes).
+ Remove from the heat and set aside until needed. If it has firmed up too much by the time you come to spread, reheat it briefly.

ASSEMBLING THE BABKA

+ Line a 900g loaf tin (23 x 13 x 7.5cm) with baking paper. Prepare a large sheet of baking paper and dust it generously with millet flour.
+ Turn out the chilled dough on to the floured baking paper, dust it generously with millet flour, and roll it out into a 28 x 38cm rectangle. Turn the rectangle so that a short end is closest to you.
+ Spoon the filling over the rolled-out dough, and spread it out with an offset spatula all the way to the edges, in an even, thin layer.
+ Starting at the short end farthest away from you, roll up the babka towards you. You should finish with a 28cm log. (If the dough feels too soft to cut at this point, chill it for 15–30 minutes before continuing.)
+ Turn the rolled-up log seam side downwards, and use a piece of floss or baker's thread to cut it in half along its length, all the way down to the

work surface, to create 2 long strands. ② If you don't manage to cut the bottom few layers with the thread, use a lightly oiled knife to finish.

+ Turn each half so that the cut sides face upwards and twist the two strands around each other. Transfer the babka into the lined loaf tin. (If you're not confident you'll be able to transfer the babka on its own, cut the baking paper to size and use it to lift the babka into the tin.)
+ Lightly cover the loaf tin with cling film (to prevent the surface of the babka from drying out) and prove in a warm place for about 45–60 minutes or until approximately doubled in volume.
+ Adjust the oven shelf to the middle position and preheat the oven to 180°C.
+ For the egg wash, whisk together the egg and milk. Once the babka has proved, gently brush it all over with the wash.
+ Bake for about 1 hour 10 minutes–1 hour 15 minutes or until golden brown on top and an inserted toothpick comes out without any raw or half-baked dough attached. (Note that the toothpick will have traces of the chocolate filling on it.) If the top browns before the babka is ready, cover with foil (shiny side up) and bake until done.

SUGAR SYRUP GLAZE

+ In a small saucepan, cook the sugar and water over a medium–high heat until all the sugar has dissolved and the mixture forms a runny syrup.
+ As soon as the babka is out of the oven, brush it with the sugar syrup.
+ Allow the babka to cool in the loaf tin for 15–20 minutes, then remove from the tin and place on a wire rack to cool completely.

STORAGE

Best on the day. Or store it in an airtight container until the following day, then reheat it in the microwave for 20–30 seconds to serve.

Baker's percentage (b%) and ratios for chocolate babka

QUANTITY NAME	VALUE
hydration	74 b%
starch	50 b%
binders	5 b%
psyllium:xanthan ratio	2:1

PHOTOGRAPHY OVERLEAF ⟶

NOTES

① *In most wheat-based babka recipes, the filling contains a lot more butter and little-to-no milk. I don't recommend using such a butter-heavy filling with the gluten-free babka, as the bottom layers of the babka absorb the melted butter during baking, making them dense in texture.*

② *Because the dough is fairly soft and sticky, I really recommend you use a long, thin piece of thread or floss instead of a sharp knife or bench scraper to halve the rolled-up dough. It will give a much cleaner cut, with well-defined layers of dough and chocolate filling.*

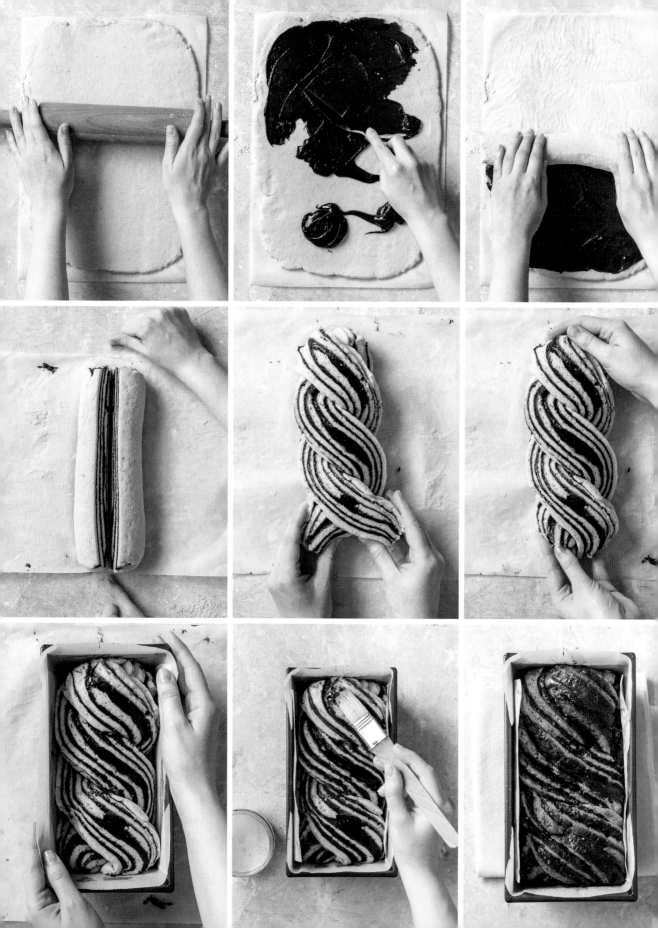

DOUGHNUTS

MAKES 15
PREP TIME 1 hour 30 mins
RISE TIME 1 hour 30 mins
CHILL TIME 30 mins
FRY TIME 5 x 5 mins

We're talking proper fried doughnuts here. The ones with a crisp, golden-brown crust and an impossibly fluffy interior. The secret behind such amazing doughnuts lies in the perfect psyllium-husk:xanthan-gum ratio; a brief refrigeration after the first rise, which sets the butter and makes the dough easier to handle; and a lower oil temperature to ensure the doughnuts cook through before the crust becomes too dark or crunchy. Because I have a weakness for icings and toppings, I have given three to choose from, all equally delicious.

BRIOCHE DOUGH

8g active dried yeast
50g caster sugar
90g whole milk, warmed
10g psyllium husk
140g warm water
155g tapioca starch
130g millet flour, plus flouring
25g sorghum flour
5g xanthan gum
5g salt
1 egg, at room temperature
35g unsalted butter, softened
sunflower oil, for frying

SIMPLE SUGAR ICING
(FOR 5 DOUGHNUTS)

150g icing sugar
2½–3½ tablespoons whole milk, at room temperature
½ teaspoon vanilla bean paste

CINNAMON TOPPING
(FOR 5 DOUGHNUTS)

50g caster sugar
½–1 teaspoon ground cinnamon

CHOCOLATE GLAZE
(FOR 5 DOUGHNUTS)

45g dark chocolate (about 60% cocoa solids), melted
15g cocoa powder
75g icing sugar
3–3½ tablespoons whole milk, at room temperature
chocolate sprinkles, to decorate

BRIOCHE DOUGH

+ In a small bowl, mix together the yeast, 10g of the sugar and the warm milk. Set aside for 10–15 minutes or until the mixture starts frothing.
+ In a separate bowl, mix together the psyllium husk and water. After about 15–30 seconds, a gel will form.
+ In a large bowl, mix together the tapioca starch, millet flour, sorghum flour, the remaining 40g sugar, and the xanthan gum and salt.
+ Add the yeast mixture, psyllium gel and egg. Using a stand mixer fitted with a dough hook or by hand, knead everything together until you get a soft and sticky dough (about 5–10 minutes).
+ Add the softened butter and knead for a further 5 minutes until you have incorporated all the butter and the dough is very soft and sticky.
+ Transfer the dough to an oiled container and allow to rise in a warm place for about 1 hour–1 hour 30 minutes or until doubled in volume.
+ Once the dough has risen, refrigerate for 30–60 minutes.

SHAPING + FRYING THE DOUGHNUTS ①

+ Cut out 15 squares of baking paper, each 9cm.
+ Turn out the chilled dough on to a lightly millet-floured surface and roll it out to about 1cm thick.
+ Dip a 7.5cm round cutter into a little millet flour and cut out circles of dough. Then, use a floured 2.5–3cm round cutter to cut out a hole in the middle of each doughnut (a large piping nozzle turned upside down works too). Re-roll the trimmings until you have 15 doughnuts.
+ Place each doughnut on a square of baking paper and transfer them all to a baking sheet. Cover loosely with cling film and prove in a warm place for about 30 minutes or until noticeably puffed up – they should increase in volume by about half.
+ While the doughnuts are proving, fill a large saucepan about ⅔ full with sunflower oil (or another neutral-tasting oil with a high smoke point) and heat it to 160–165°C. ②
+ Carefully lower 3 of the proved doughnuts at a time, each with their square of baking paper, into the hot oil. ③ After about 30 seconds, remove the squares of baking paper with kitchen tongs.
+ Fry the doughnuts for about 2½–3 minutes on each side, then transfer

to a plate lined with kitchen paper. Cover with another sheet of kitchen paper and a piece of foil to trap the steam and make the crust softer.
+ Repeat with the remaining doughnuts, making sure that the oil temperature remains at a constant 160–165°C throughout. Allow all the doughnuts to cool until warm.

SIMPLE SUGAR ICING
+ Whisk together all the icing ingredients until smooth.
+ Dip 5 doughnuts into the icing, then place them on a wire rack so any excess icing can drip away.
+ Allow the icing to set and harden for about 20 minutes, then serve.

CINNAMON TOPPING
+ Mix together the sugar and cinnamon.
+ Dip 5 doughnuts into the topping, coating all sides, and serve.

CHOCOLATE GLAZE
+ In a microwave-safe bowl, whisk together the melted chocolate, cocoa powder, icing sugar and 2 tablespoons of milk to a thick paste. Microwave until warm, then add the remaining milk, ½ tablespoon at a time, to a thick but runny consistency that coats the back of a spoon. You shouldn't need more than 3½ tablespoons of milk – if the icing gets too thick, just reheat it briefly until warm.
+ Dip 5 doughnuts into the glaze, then place them on the wire rack so any excess glaze can drip away.
+ Sprinkle with chocolate sprinkles, allow the glaze to set for about 20–30 minutes, then serve.

STORAGE
Best within a few hours of frying, but you can keep the undecorated doughnuts in an airtight container for up to 1 day, then reheat in the microwave for 15–20 seconds before finishing and serving.

Baker's percentage (b%) and ratios for doughnuts

QUANTITY NAME	VALUE
hydration	74 b%
starch	50 b%
binders	5 b%
psyllium:xanthan ratio	2:1

NOTES

① *Although the method describes how to shape ring doughnuts, the recipe also works well for filled doughnuts. Just omit cutting a hole in the doughnuts, then prove and fry as directed. Then, use a piping bag fitted with a round piping nozzle to fill the warm or cooled doughnuts with your filling of choice.*

② *This oil temperature is slightly lower than the norm (usually in the 180–190°C range) because gluten-free doughnuts need slightly longer to cook through. Frying them at a higher temperature would give a much darker and crunchier crust in the time it takes for the middle to be done.*

③ *Gluten-free doughnuts are slightly more fragile than their wheat-based counterparts – placing them on squares of baking paper allows you to transfer them into the hot oil easily, without deflating the dough or destroying their circular shape.*

DRIED FRUIT SODA BREAD

MAKES 1 loaf
PREP TIME 30 mins
BAKE TIME 1 hour 10 mins

50g sultanas or raisins
hot water, for soaking
9g psyllium husk
90g water
185g millet flour, plus extra for
    flouring the surface
185g tapioca starch
40g oat flour
40g sorghum flour
40g caster sugar
6g xanthan gum
6g salt
1½ teaspoons baking powder
1½ teaspoons bicarbonate of soda
360g buttermilk, at room
    temperature
1 tablespoon whole milk
2 teaspoons granulated sugar

NOTES

① *Hydrating the sultanas makes
them juicy, plump and more
pleasant to eat. The moisture in
the bread isn't enough to hydrate
them to the same extent.*

② *Making the dough fairly flat and
cutting a fairly deep cross into
it isn't just for aesthetics. Both
actions help the bread to bake
through before the crust gets
too hard, which is especially
relevant with gluten-free flours.*

③ *The milk prevents the crust from
getting too hard, while the sugar
gives it a wonderful flavour and
appearance.*

Although soda bread doesn't rely on yeast for its rise, but rather on chemical leavening agents, the basic principle behind the rise is the same in both doughs: the dough traps a gas (this time generated from the bicarbonate of soda), which heats up and expands, causing the bread to expand along with it. That's why it's necessary to add both psyllium husk and xanthan gum. And while most soda breads use only bicarbonate of soda, I like my version to use baking powder too, for that extra help in getting a soft crumb. I've specifically chosen this combination of flours to mimic the flavour of wheat soda bread. They certainly get the job done.

+ Adjust the oven shelf to the lower-middle position, preheat the oven to 180°C and line a baking sheet with baking paper.
+ Place the sultanas or raisins into a small bowl, cover them with hot water and leave for about 5–10 minutes to hydrate and soften. Pour the water away and gently squeeze out any excess moisture from the fruit. Set aside until needed. ①
+ In a separate small bowl, mix together the psyllium husk and the 90g water. After about 15–30 seconds, a gel will form.
+ In a large bowl, mix together the millet flour, tapioca starch, oat flour, sorghum flour, caster sugar, xanthan gum, salt, baking powder and bicarbonate of soda.
+ Add the psyllium gel and buttermilk, and (by hand) knead to a shaggy dough. When the dough starts coming together, add the sultanas or raisins and knead to evenly distribute. The final dough shouldn't have any patches of dry flour, but it doesn't have to be completely smooth.
+ Turn out the dough on to a lightly floured surface and knead it gently while forming it into a ball. Transfer the dough on to the lined baking sheet and pat it down into a disc about 6.5cm thick. Using a sharp knife, cut a cross, about 1cm deep, in the top. ② Gently brush the bread with the milk and sprinkle with the granulated sugar. ③
+ Bake for about 1 hour 10 minutes or until the loaf is golden brown, a toothpick inserted into the middle comes out clean (with no unbaked dough attached) and the bread sounds hollow when tapped on the base. If the top of the bread browns too quickly, cover it with foil (shiny side up) and bake until done.
+ Transfer the baked soda bread to a wire rack to cool completely. The weight of the loaf straight out of the oven should be about or below 930g (about 11% weight loss), and it will continue losing weight while cooling to reach a final weight of about 900g (about 14% weight loss).

STORAGE

3–4 days in an airtight container in a cool, dry place. Although the bread is best toasted on days 3 and 4.

BREAKFAST
+
TEATIME
TREATS

From the fluffiest American-style pancakes drenched in maple syrup to a chocolate chip banana bread with the most gloriously caramelised edges, this chapter is all about low-effort comfort food you might enjoy for a lazy breakfast or as a teatime treat.

These are the kinds of recipe that don't require extraordinary amounts of prep or advance planning. They are the kinds that you can throw together, pop into the oven and leave alone to do their thing. Or, they are the kinds that you mix together and make on the hob in fewer than 30 minutes.

Above all, they are the good-old favourites we return to time and again – among them delicate, lacy French crêpes, buttery scones and a simple (yet terribly delicious) Victoria sponge cake.

VANILLA FRENCH CRÊPES

MAKES 7
PREP TIME 5 mins
COOK TIME 10 mins

120g gluten-free flour blend
¼ teaspoon xanthan gum
1 tablespoon caster sugar ①
¼ teaspoon salt
2 eggs, at room temperature
1 tablespoon melted unsalted
    butter (optional) ②, plus extra
    for frying
300g whole milk, at room
    temperature
maple syrup, berries, nuts or
    other toppings, to serve

NOTES

① *As well as making the crêpes
    just sweet enough, sugar gives
    the crêpes their characteristic
    browning. This browning occurs
    because of the 'Maillard reaction'
    between amino acids (proteins)
    and sugars in the presence of
    heat. It also forms new, delicious
    aromas and flavours.*

② *Adding butter is optional but I
    recommend it because the butter
    ensures that the crêpes stay soft
    and delicate for longer, as well
    as giving them a richer flavour.*

③ *This method of first adding
    only part of the wet to the dry
    ingredients reduces clump
    formation, creating an initial
    loose paste that you dilute with
    the remaining wet ingredients,
    making it far easier to get a
    smooth, homogeneous batter.*

There are few things as delightful as a plateful of lacy French crêpes that you can serve up in only 15 minutes. They're thin, delicately soft with a bit of crunch at the edge, like they've magically popped into your kitchen all the way from a quaint Parisian crêperie. The secret to perfect crêpes lies in a very hot frying pan. Make sure to preheat your pan before you start.

+   In a large bowl, whisk together the gluten-free flour blend, xanthan gum, sugar and salt.
+   Add the eggs, melted butter and about ⅓ of the milk. ③ Whisk well until you get a smooth, thick batter.
+   Gradually add the remaining milk, whisking continuously. If your finished batter has any lumps, pass it through a fine sieve.
+   Heat up a skillet or frying pan until a drop of water sizzles and bounces. Lightly butter the pan and wipe away any excess with kitchen paper to prevent the butter from burning. Add about a ladleful of batter. As you pour, swirl the pan to get an even, thin coating.
+   Cook over a medium–high heat for about 45 seconds or until the crêpe loosens from the pan and is golden brown underneath.
+   Flip and cook for a further 30–45 seconds on the other side until dark brown freckles appear, then place it on a plate or wire rack while you repeat with the remaining batter until you have 7 crêpes.
+   Serve immediately with toppings of your choice.

STORAGE

The crêpes will stay soft for a few hours if wrapped in cling film, but eat them on the same day.

PUT A TWIST ON IT

Chocolate crêpes – Add 1–2 tablespoons of Dutch processed cocoa powder to the dry ingredients, and increase the amount of sugar to 2 tablespoons.

FLUFFY AMERICAN PANCAKES

MAKES 6
PREP TIME 10 mins
COOK TIME 12 mins

There's nothing like an inviting stack of fluffy pancakes, topped with berries and dripping with maple syrup on a lazy Sunday morning. Feel free to adapt this simple recipe further – toss a handful of blueberries into the batter, add cocoa powder, chocolate chips or even melted chocolate for a morning chocolate fix, stuff them with hazelnut spread – the options are as endless as they are delicious.

130g gluten-free flour blend
2 tablespoons caster sugar
2 teaspoons baking powder
½ teaspoon xanthan gum
¼ teaspoon salt
200g whole milk, at room temperature
2 tablespoons melted butter, plus extra for frying
1 egg, at room temperature
maple syrup, fruit or other toppings, to serve

+ Sift together the gluten-free flour blend, caster sugar, baking powder, xanthan gum and salt. Add the milk, melted butter and egg, and whisk until you get a smooth, thick batter. ①
+ Heat a frying pan over a medium–high heat until a drop of water sizzles and bounces. Lightly butter the pan and drop about a ladleful of the pancake batter on to it, spreading it around gently with the back of a ladle or spoon to make a circular shape about 15cm in diameter.
+ Cook the pancake for about 2 minutes until the edges are set (not sticky to the touch), bubbles appear on the surface and the pancake is golden brown underneath. Flip the pancake and cook for a further 2 minutes on the other side.
+ Once done, transfer the pancake to a plate lined with kitchen paper to absorb any condensation that forms under the hot pancake. Repeat until all the batter is used up and you have 6 pancakes. (Depending on the size of your frying pan, cook the remainder in batches of perhaps 2 or 3 pancakes at a time. Butter the pan between batches.)
+ Serve warm immediately after cooking, topped with fresh fruit or other toppings and drizzled with maple syrup, if you wish.

NOTE

① *The thick consistency is key to getting tall, fluffy pancakes. A runny pancake batter causes the pancakes to spread, making them less American in style and more closely resembling soft crêpes. For that reason, you're adding less milk in ratio to flour than you would for crêpes.*

STORAGE
Best as you make them, but they stay moist and fluffy for a few hours afterwards.

CHOCOLATE CHIP BANANA BREAD

SERVES 10
PREP TIME 15 mins
BAKE TIME 1 hour

3 bananas (about 330g, peeled
  weight), mashed
3 teaspoons lemon juice
150g unsalted butter, melted
  and cooled
150g light brown soft sugar
2 eggs, at room temperature
1 teaspoon vanilla bean paste
225g gluten-free flour blend
50g almond flour
1½ teaspoons baking powder
¾ teaspoon bicarbonate of soda
¼ teaspoon xanthan gum
¼ teaspoon salt
½ teaspoon ground cinnamon
¼ teaspoon ground nutmeg
100–175g dark or milk chocolate
  chips

NOTE

① *The colour of the loaf tin
  affects the extent to which
  the crust and edges of the
  banana bread caramelise. For
  gently caramelised edges, use
  a light-coloured tin; for a darker
  crust, use a darker tin, which
  leads to faster and more intense
  caramelisation. Both work
  well in this loaf, but for me it
  has to be the darker.*

From the caramelised crust to the fluffy, moist interior studded with chocolate chips, this banana bread is the very definition of cosy comfort food. It takes only about 15 minutes to put together – although waiting for an hour for it to bake (and then at least half an hour for it to cool) while your kitchen fills with mouthwatering smells… that's certainly an exercise in willpower.

+ Adjust the oven shelf to the middle position, preheat the oven to 180°C and line a 900g loaf tin (23 x 13 x 7.5cm) with baking paper. ①
+ Mix the bananas with the lemon juice (to prevent oxidation and browning), and set aside until needed.
+ In a bowl, whisk together the melted butter and the sugar until pale and fluffy. (No need to use a stand or electric hand mixer – a balloon whisk aerates the batter more than enough.)
+ Add the mashed bananas, eggs and vanilla paste, and whisk until well combined.
+ Sift in the gluten-free flour blend, almond flour, baking powder, bicarbonate of soda, xanthan gum, salt, cinnamon and nutmeg, and mix well until you get a smooth batter with no flour clumps.
+ Reserve some of the chocolate chips for sprinkling, then mix the remainder into the batter until evenly distributed.
+ Transfer the batter to the loaf tin, smooth out the top, sprinkle with the reserved chocolate chips and bake for about 1 hour or until well risen, golden brown on top and an inserted toothpick comes out clean or with a few stray crumbs attached. If the banana bread browns too quickly, cover with foil (shiny side up) and bake until done.
+ Allow to cool in the loaf tin for 15–30 minutes, then transfer to a wire rack to cool completely (although it's absolutely delicious still slightly warm, with the chocolate chips still melted).

STORAGE
3–4 days in a closed container at room temperature.

PUT A TWIST ON IT
Double chocolate banana bread – Add a few tablespoons of Dutch processed cocoa powder into the batter. Or for a nuttier, chocolate twist, swirl chocolate hazelnut spread into the batter or add chopped, toasted pecans or hazelnuts.

EXTRA-FLAKY SCONES

MAKES 9
PREP TIME 30 mins
BAKE TIME 18 mins

240g gluten-free flour blend
2 teaspoons xanthan gum
1½ teaspoons baking powder
½ teaspoon salt
150g unsalted butter, cubed
    and chilled
35g caster sugar
80g whole milk, cold, plus an extra
    1 tablespoon for the egg wash
3 tablespoons melted butter
1 egg
1 tablespoon granulated sugar
jam of choice, to serve
double cream, whipped to soft
    peaks, to serve

NOTES

① *If the dough feels too soft, it's
   probably because the butter
   warmed up too much while
   you were kneading. Chill it for
   about 15 minutes before rolling.
   Similarly, chill it at any point
   during the rolling and folding
   process if it feels too soft.*

② *I've found scones to be highly
   sensitive to the size of the oven.
   I've optimised the baking times
   here for a 60–70-litre oven. If you
   have a smaller (say 50-litre) oven,
   where the heat source is closer
   to the scones, shorten the time
   by up to 5 minutes for equally
   perfect results.*

The method of making these scones, where you brush the rolled-out dough with melted butter and do a series of two folds, is reminiscent of the laminating process for making (rough) puff pastry – only much easier and quicker. The result is scones with more defined butter–dough layers, and consequently a more pronounced flake. The lamination, along with the baking powder, helps them reach greater heights during baking, while the final fold in half gives a natural crease that makes it easier to tear and fill the baked scones. Finally, a baking temperature of 190°C leads to a rapid and dramatic evaporation of water in the butter, which I find results in a delicate scone that is still able to hold the filling without breaking.

+ Adjust the oven shelf to the middle position, preheat the oven to 190°C and line a baking sheet with baking paper.
+ In a large bowl, sift together the gluten-free flour blend, xanthan gum, baking powder and salt. Add the chilled butter and work it into the flour until the largest butter pieces are about the size of peas. Add the sugar and mix briefly to combine.
+ Add the milk and mix until you get a shaggy dough. If necessary, knead the dough gently until it comes together in a ball. ①
+ Roll out the dough on a lightly floured work surface to a 15 x 30cm rectangle. Turn the dough so that one of the short ends is closest to you, then brush the top of the rectangle with half the melted butter. Fold the dough in thirds as you would an A4 letter, then rotate by 90 degrees so that the open (short) ends are closest and farthest from you. Roll again to a 15 x 30cm rectangle. Brush the top half with butter and fold the dough in half. You should get a 15cm square about 2–2.5cm thick. If necessary, roll out the dough to this thickness.
+ Divide the dough into 9 smaller, about 5cm squares and transfer them to the lined baking sheet.
+ Combine the egg and the 1 tablespoon of milk, whisking lightly, to create a wash. Use this to brush the tops of the dough squares, then sprinkle with granulated sugar. (Ensure none of the egg wash drips down the sides, which can lead to an uneven rise and lopsided results.)
+ Bake for about 18–20 minutes or until the scones have risen and are deep golden brown. ② The scones should be about 3–3.5cm tall.
+ Serve warm or cold, topped with a jam of choice and whipped cream.

STORAGE

Best on the day. Or, keep the scones in a closed container for the following day, then reheat in the microwave for 10–15 seconds to serve.

VICTORIA SPONGE CAKE

SERVES 12
PREP TIME 45 mins
BAKE TIME 25 mins

Simplicity at its best. Vanilla-scented, fluffy, buttery sponges sandwiched together with a generous layer of whipped cream and sweet strawberry jam – what's not to love? I've used 5 eggs for the two sponges to optimise the cake:icing ratio for ultimate enjoyment. With just a final dusting of icing sugar, this classic British teatime treat beckons you to help yourself to a slice.

225g gluten-free flour blend
50g almond flour
3 teaspoons baking powder
¾ teaspoon xanthan gum
275g unsalted butter, softened
275g caster sugar
5 eggs, at room temperature
1 teaspoon vanilla bean paste
250g double cream, chilled
50g icing sugar, plus extra for dusting
150–175g strawberry jam

NOTE

① *This all-in-one method, aside from being easier and more time efficient, produces slightly sturdier sponges than the more common standard creaming method. The cakes are less likely to collapse, have a moister mouthfeel and are better suited to a stacked layer cake such as this.*

+ Adjust the oven shelf to the middle position, preheat the oven to 180°C and line two 20cm round cake tins with baking paper.
+ Using a stand mixer with the paddle attachment or a hand mixer fitted with the double beaters, or by hand using a balloon whisk, combine the gluten-free flour blend, almond flour, baking powder, xanthan gum, butter, sugar, eggs and vanilla paste in a large bowl until smooth and thick, with no flour or butter clumps. ①
+ Divide the batter equally between the lined cake tins and smooth out the tops. Bake for about 25–30 minutes or until an inserted toothpick comes out clean. Do not open the oven door during the first 20 minutes, as doing so can cause the sponges to collapse.
+ Allow the sponges to cool in the baking tins for about 10 minutes, then turn them out on to a wire rack to cool completely.
+ Whip the double cream and icing sugar together until soft peaks form. Spread the strawberry jam on top of one of the sponges, then spoon over the whipped cream and gently spread it out in an even layer. Top with the other sponge and finish with a dusting of icing sugar.

STORAGE
1–2 days in a closed container in a cool, dry place.

CHOCOLATE-GLAZED STRAWBERRY BAKED DOUGHNUTS

MAKES 6
PREP TIME 30 mins
BAKE TIME 12 mins

These aren't your usual baked doughnuts – these are filled baked doughnuts, with a strawberry jam centre, glazed with chocolate icing. Filling the doughnuts is simple: partially fill the cavities with batter, pipe a ring of strawberry jam and top with the remaining batter. Finding that jam centre in what is already a delicious chocolate-glazed doughnut will definitely put a smile on your face.

BAKED DOUGHNUTS

100g gluten-free flour blend
20g almond flour
50g light brown soft sugar
1 teaspoon baking powder
¼ teaspoon xanthan gum
¼ teaspoon salt
1 egg, at room temperature
50g unsalted butter, melted and cooled, plus extra for greasing
40g whole milk, at room temperature
40g full-fat plain yoghurt, at room temperature
½ teaspoon vanilla bean paste
6 teaspoons strawberry jam
freeze-dried strawberries, gently crushed, to decorate

CHOCOLATE GLAZE

45g dark chocolate (about 60% cocoa solids), melted
15g Dutch processed cocoa powder
75g icing sugar
3–3½ tablespoons whole milk, at room temperature

NOTE

① *If any of the holes have partly closed during baking, while the doughnuts are hot, use an apple corer or the large end of a piping nozzle to cut away any excess.*

BAKED DOUGHNUTS

+ Adjust the oven shelf to the middle position, preheat the oven to 180°C and lightly butter a 6-hole doughnut tin with 9cm doughnut cavities.
+ In a bowl, whisk together the gluten-free flour blend, almond flour, sugar, baking powder, xanthan gum and salt.
+ In a separate bowl, whisk together the egg, melted butter, milk, yoghurt and vanilla paste. Add the wet to the dry ingredients and combine to a smooth batter with no flour clumps.
+ Transfer the batter to a piping bag fitted with a large round nozzle. Stir the strawberry jam to loosen it, and spoon it into a separate piping bag fitted with a small round nozzle.
+ Pipe a layer of doughnut batter into the tin cavities, filling each about ⅓ full. Use a spoon or offset spatula to make a groove the centre of the batter. Pipe about 1 teaspoon of the jam into the groove, then pipe more batter on top, so that each cavity is about ⅔–¾ full. Use the spoon or spatula to gently spread the top layer of batter so that no jam is visible.
+ Bake for about 12–14 minutes or until the doughnuts have risen, are golden brown on top and an inserted toothpick comes out clean. Turn the doughnuts out on to a wire rack to cool completely. ①

CHOCOLATE GLAZE

+ In a microwave-safe bowl or in a small pan, whisk the melted chocolate, cocoa powder, icing sugar and 2 tablespoons of milk to a thick paste. Heat in a microwave or over a medium heat until warm, then add the remaining milk, ½ tablespoon at a time, to a thick, runny consistency that coats the back of a spoon – you shouldn't need more than 3½ tablespoons of milk in total. If the icing gets too thick, reheat it briefly.
+ Dip the cooled doughnuts into the icing, then place them on the wire rack so any excess icing can drip away.
+ Sprinkle with the freeze-dried strawberries, and allow the icing to set for at least 20–30 minutes before serving.

STORAGE

Best on the day. Or, keep the doughnuts in a closed container until the following day and reheat in the microwave for 5–10 seconds to serve.

CLASSIC POUND CAKE

SERVES 10–12
PREP TIME 30 mins
BAKE TIME 1 hour

250g unsalted butter, softened
250g caster sugar
5 eggs, at room temperature
220g gluten-free flour blend
¾ teaspoon xanthan gum

NOTES

① *For gently caramelised edges (my preference), it's best to use a light-coloured loaf tin. A darker tin leads to faster and more intense caramelisation, giving a thicker, dark brown crust.*

② *Because the pound cake relies on incorporated air for its rise, it's crucial that you maintain the butter emulsion throughout as you add the eggs (that is: you don't want the mixture to split and curdle) – that's why you add the eggs in very small batches.*

I love a good pound cake. Not just because every bite is sweet, buttery deliciousness, but also because the science behind it is incredibly simple and yet also incredibly effective. This cake rises beautifully in the oven, producing a domed top with a pretty golden crack down the middle, and yet it has no raising agents (such as baking powder), and it doesn't rely on whipped eggs for its lift. The rise is all down to the aeration of the batter, which means that the mixing method is especially important: creaming the butter and sugar until pale and fluffy, adding the eggs in about ten batches and mixing them well in between, sifting the dry ingredients… all those actions are about aerating the batter and maintaining the emulsion of the butter, so that the mixture doesn't curdle as you're adding the eggs.

+ Adjust the oven shelf to the middle position, preheat the oven to 180°C and line a 900g loaf tin (23 x 13 x 7.5cm) with baking paper. ①
+ Using a stand mixer with the paddle attachment or a hand mixer fitted with the double beaters, cream the butter for 1–2 minutes. Add the caster sugar and beat for a further 3–5 minutes until pale and fluffy.
+ In a separate bowl, whisk together the eggs, then add them slowly, in about 10–12 batches, to the butter–sugar mixture. ② Thoroughly mix the batter after each addition, starting the mixer on low speed and finishing on high.
+ Once you have incorporated all the eggs, sift in the gluten-free flour blend and xanthan gum. Mix until combined, again starting the mixer on low speed and ending on high for 10–15 seconds.
+ Transfer half the batter to the lined tin. Use a small, offset spatula to spread it out into an even layer, then add the remaining batter on top, spreading that out evenly too. To get a more even crack along the top of the cake, use a sharp knife or toothpick to gently draw a line lengthways, about 1mm deep, down the middle of the batter.
+ Bake for 1 hour–1 hour 10 minutes until risen, cracked in the middle, a golden brown colour on top and an inserted toothpick comes out clean. If the cake starts browning too quickly, cover with foil (shiny side up) and bake until done.
+ Remove from the oven and allow to cool in the tin for 10–15 minutes, then transfer to a wire rack to cool completely.

STORAGE

Best on the day. Or, keep the cake in a closed container for 3–4 days. On days 3 and 4, it's best toasted on a griddle pan or in a frying pan on a bit of butter. You can use any leftovers in bread pudding.

AROUND THE WORLD

A confession: this chapter exists because I simply couldn't imagine the book without a recipe for gluten-free choux pastry (a recipe I'm incredibly proud of), but it didn't quite fit among the cakes, cupcakes and cookies. And yet, from such random, irreverent beginnings has emerged a chapter of delicious favourites from around the world.

There are the coffee choux buns (the very definition of indulgent), the hypnotising flaky layers of millefeuille, the absolute legend that is coffee-soaked tiramisù (with from-scratch, real-deal ladyfinger biscuits) and the flavour explosion of well-made linzer cookies.

The science behind each recipe is incredibly interesting, but there's no common scientific thread. So, we'll take each recipe as we go, one by one looking at the chemistry and interactions of the ingredients.

Some of these recipes may sound daunting to attempt with wheat flour, let alone within the constraints of gluten-free baking. But, behind the fancy-sounding names and the beautiful exteriors, these bakes are no more difficult nor impossible than the average cake. And, yes, that includes the éclairs. (Plus, the results are so worth it.)

THE PERFECT CHOUX PASTRY

MAKES 1 quantity
PREP TIME 30 mins
BAKE TIME see individual
recipes (pages 338 and 342)

100g water
100g whole milk
90g unsalted butter, cubed
10g caster sugar
¼ teaspoon salt
100g gluten-free flour blend
2 teaspoons xanthan gum
4 eggs, at room temperature

NOTES

① *This method of adding hot liquid to the flour ensures a silky-smooth choux batter. The wheat-based method of adding all the flour at once to the boiling-hot water–milk mixture can (with gluten-free flours) often result in flour clumps. These, in turn, can cause an uncontrolled rise of the choux, resulting in cracks and irregular shapes.*

② *The exact amount of egg you need depends on a combination of factors, such as how much liquid has evaporated during the panade cooking, the exact gluten-free flour blend you've used, and much besides. The more moisture there is in the panade, the fewer eggs you'll need to add (and vice versa).*

This recipe will guide you, step by step, to the perfect gluten-free choux batter, ready to be turned into cream puffs or French éclairs (see pages 338 and 342). If you follow it to the letter, you'll end up with a silky-smooth batter that expands in the oven to produce a golden, light, crisp, hollow choux shell, just crying out to be filled with a decadent, flavoured cream. Because gluten-free flours have a greater water-absorption capacity than wheat flour (so they're quicker to form lumps in the presence of liquid), there are a few extra steps to keep your choux smooth at every stage. But even with that in mind, making gluten-free choux pastry really isn't very difficult at all.

+ Adjust the oven shelf to the middle position, preheat the oven to 230°C and line a baking tray or sheet with baking paper.
+ Put the water, milk, butter, sugar and salt in a saucepan over a medium–high heat and bring to the boil. By the time the mixture is boiling, the butter should be completely melted and the salt and sugar completely dissolved.
+ In the bowl of a stand mixer with the paddle attachment or using a hand mixer fitted with the double beaters, mix together the gluten-free flour blend and xanthan gum. Whisking continuously, slowly add the boiling-hot water–milk mixture until you have a thick paste. ① Beat the mixture until it's completely smooth with no flour clumps.
+ Transfer the batter back into the saucepan and cook over a medium heat for 2–3 minutes until the dough comes away from the sides and starts forming a ball. Determining when the dough (called a 'panade') is ready depends on the type of saucepan: in a stainless-steel saucepan, look for a thin film of dough on the bottom; in a non-stick saucepan, look for oil droplets forming on the bottom. Remove the pan from the heat and transfer the panade back into a bowl.
+ Mix the panade for 30–60 seconds, allowing it to cool slightly (so that in the next step it doesn't scramble the eggs) and release even more moisture in the form of steam.
+ Add the whisked eggs to the panade in five or six batches. After each addition, beat the mixture until the eggs have been completely incorporated, scraping down the sides of the bowl as necessary.
+ The final batter should be smooth and glossy, and of a thick but pipeable consistency. To test if the batter is ready, use a spatula to stir the dough, then as you slowly lift the spatula, the batter should form a V-shape at the end. To get to this point, you might not need all the egg, or you might need to add an extra ½ egg (whisk it before you halve it), so test frequently. ②
+ Transfer the choux batter to a piping bag fitted with a large round or star nozzle, and pipe choux puffs (see page 338) or éclairs (see page 342) and bake as directed in the recipes.

COFFEE CREAM PUFFS (CHOUX AU CRAQUELIN)

MAKES 24
PREP TIME 1 hour 30 minutes
(including the choux batter)
CHILL TIME 30 mins
BAKE TIME 2 x 40 mins
COOK TIME 5 mins

CRAQUELIN
90g gluten-free flour blend
¼ teaspoon xanthan gum
90g light brown soft sugar
70g unsalted butter, softened

CHOUX PUFFS
1 quantity of **choux batter**
  (see page 336)
24 coffee beans, to decorate

COFFEE PASTRY CREAM
250g whole milk
4–5 teaspoons instant coffee
  granules
3 egg yolks
75g caster sugar
25g cornflour
25g unsalted butter

VANILLA WHIPPED CREAM
600g double cream, chilled
200g icing sugar, sifted
1 teaspoon vanilla bean paste

There are desserts you remember for ever. And, I promise you, these coffee-cream puffs will have you daydreaming about them days, weeks and months later. From the crisp choux pastry, with the added crunch of the craquelin topping, to the luscious coffee-cream filling and the swirl of vanilla whipped cream, these elegant little treats are as delicious to eat as they are beautiful to behold.

CRAQUELIN ①
+ Mix together all the craquelin ingredients until you get a smooth dough. Then, roll out the dough between two pieces of baking paper until it's about 1mm thick.
+ Refrigerate for at least 30 minutes or until solid to the touch.

CHOUX PUFFS
+ Adjust the oven shelf to the middle position, preheat the oven to 230°C and line two baking sheets with baking paper. Prepare the choux batter (see page 336).
+ Transfer the batter to a piping bag fitted with a large round piping nozzle and, holding the piping bag vertically upright, pipe 12 rounds of choux batter about 4cm wide and 4cm tall on to one of the lined baking sheets, spacing them about 3.5–4cm apart.
+ Take the craquelin out of the fridge and peel off the top sheet of baking paper. Turn the craquelin upside down and peel off the second sheet. Using a 5cm round cutter, cut out discs of the craquelin and place each disc on top of a piped choux round, pressing down gently.
+ Reduce the oven temperature to 170°C and place the unbaked puffs in the oven. Bake for 20 minutes, then open the oven door slightly to let out the steam. Close the door and bake the puffs for a further 20 minutes or until they are a dark golden brown, feel crisp to the touch and sound hollow when tapped.
+ Remove the puffs from the oven, make a small hole in the side of each with a toothpick (to release the steam), then transfer to a wire rack to cool completely.
+ Repeat for the remaining batter to make another 12 choux buns, giving you 24 in total.

COFFEE PASTRY CREAM

+ In a saucepan, mix together the milk and coffee, and bring to the boil.
+ Meanwhile, whisk together the egg yolks and sugar until pale. Add the cornflour and mix until combined.
+ Slowly pour the hot milk into the egg mixture, whisking vigorously all the time.
+ Return the mixture to the saucepan, and cook over a medium–high heat, stirring continuously, until thickened (about 5 minutes).
+ Remove from the heat and add in the butter. Allow to cool, stirring occasionally to prevent a skin forming.

VANILLA WHIPPED CREAM

+ In a large bowl, whip together the double cream, icing sugar and vanilla paste until the mixture only just begins to hold stiff peaks.

ASSEMBLING THE CREAM PUFFS

+ Gently fold ⅓ of the vanilla whipped cream into the coffee cream. Transfer the mixture to a piping bag fitted with an open star nozzle (use any size that suits).
+ With a sharp, serrated knife, cut off the top third of a choux bun. Pipe the coffee cream into the bottom choux cavity until it reaches the brim.
+ Transfer the remaining vanilla whipped cream into a separate piping bag fitted with an open star nozzle and pipe a swirl on top of the coffee cream layer, and replace the lid of the choux bun to cover. Repeat with the remaining puffs, coffee cream and most of the vanilla whipped cream.
+ To finish, pipe a small swirl of vanilla whipped cream on top of each filled cream puff, and decorate with a coffee bean.

STORAGE

Keep the unfilled choux buns for up to 2 days in an airtight container at room temperature. Once filled, eat within 4 hours.

NOTE

① *The craquelin serves two purposes. First, it adds a wonderful crunch to the cream puffs. Second, it controls the way in which the choux pastry expands during baking, giving the cream puffs a regular round shape. Without the craquelin, the cream puffs will have a more irregular, rustic shape.*

PHOTOGRAPHY OVERLEAF ⟶

CHOCOLATE + CARAMEL ÉCLAIRS

MAKES 16
PREP TIME 1 hour 30 minutes
(including the choux batter)
COOK TIME 10 mins
CHILL TIME 1 hour
BAKE TIME 2 x 40 mins

CARAMEL WHIPPED CREAM

100g caster sugar
600g double cream, at room
    temperature
150g icing sugar
1 teaspoon vanilla bean paste

ÉCLAIR BUNS

1 quantity of **choux batter** (see
    page 336)
1–2 tablespoons icing sugar,
    for dusting

CHOCOLATE GLAZE

170g dark chocolate (about
    60% cocoa solids), chopped
100g unsalted butter
1 tablespoon golden syrup

CARAMEL SHARDS

75g caster sugar
1 tablespoon water

Crisp, golden choux pastry filled with the most amazing caramel whipped cream, covered in a thin, shiny chocolate glaze and decorated with amber caramel shards – and you can make all of this in your kitchen with no problems whatsoever. Sure, there is a bit of extra effort required, but every decadent mouthful makes the time you've spent so very worth it.

CARAMEL WHIPPED CREAM

+ In a saucepan (preferably a stainless- steel one), cook the caster sugar over a medium–high heat until melted and caramelised to deep amber.
+ Remove the pan from the heat and slowly pour in half the double cream, stirring or whipping vigorously. If any of the sugar seizes up, return the pan to the heat and cook until dissolved. Then, add the rest of the cream and mix well.
+ Pass the caramel cream through a sieve to remove any stray pieces of solid sugar, then refrigerate for about 1–2 hours until thoroughly chilled.

ÉCLAIR BUNS

+ Adjust the oven shelf to the middle position, preheat the oven to 230°C and line two baking sheets with baking paper. Prepare the choux batter (see page 336).
+ Transfer the choux batter into a piping bag fitted with a 1cm French star piping nozzle, and, holding the piping bag at a 45° angle, pipe 8 éclairs on to one of the lined baking sheets about 2.5cm apart – they should be about 2cm wide and 10–11cm long. ① Using a finger dipped in water, smooth away any rough edges.
+ Dust the éclair buns with icing sugar, reduce the oven temperature to 170°C and place the buns in the oven. Bake for 20 minutes, then open the oven door slightly to let out the steam. Close the door and bake for a further 20 minutes or until the buns are a dark golden brown, feel crisp to the touch and sound hollow when tapped.
+ Remove the buns from the oven and make a small hole in the side of each with a toothpick (to release the steam). Cool on a wire rack for about 10 minutes, then cut in half horizontally (keep the matching halves paired up) and allow to cool completely.
+ Repeat for the remaining choux batter to make another 8 éclair buns, giving you 16 in total.

CHOCOLATE GLAZE

+ Melt the chocolate, butter and golden syrup in a heatproof bowl over a pan of simmering water (or in the microwave). Mix well to get a shiny, runny glaze. Allow to cool until it reaches 32–34°C. ②
+ Dip the top halves of the éclair buns into the chocolate glaze, leaving a 3–4mm glaze-free border around the edge. Gently shake the coated bun halves to allow any excess glaze to drip away, and use a toothpick to wipe off any excess chocolate along the edges.
+ Set aside at room temperature to allow the glaze to set.

CARAMEL SHARDS

+ Place a piece of baking paper on a baking sheet.
+ Combine the sugar and water in a saucepan (preferably stainless steel) and heat until the sugar is dark golden brown.
+ Pour the caramel on to the baking paper in an even, thin layer.
+ Allow to cool and harden, then cut into small shards with a sharp knife.

ASSEMBLING THE ÉCLAIRS

+ Add the icing sugar and the vanilla paste to the chilled caramel cream, then using a stand mixer with the whisk attachment or a hand mixer fitted with the double beaters, whip to soft peaks.
+ Transfer the whipped cream to a piping bag fitted with a rose petal nozzle (or other nozzle of choice) and pipe equal amounts on to the bottom half of each éclair bun. Top each with the matching chocolate-dipped top half, to sandwich. Decorate with caramel shards to serve.

STORAGE

Keep the unfilled éclairs for up to 2 days in an airtight container at room temperature. Once filled, eat within 4 hours.

NOTES

① *The French star nozzle gives the best éclair shape as it gives a large surface area from the outset, allowing the buns to expand easily in the oven. This reduces the likelihood of the éclair buns cracking, as might happen if you used a round nozzle. If you don't have a French star nozzle, a large open star nozzle is the next-best thing.*

② *At 32–34°C, the glaze sets with a beautiful glossy finish and is runny enough to make a thin, delicate layer on top of the éclairs.*

PHOTOGRAPHY OVERLEAF ⟶

RASPBERRY + CHOCOLATE OPERA CAKE

SERVES 6
PREP TIME 1 hour 30 minutes
BAKE TIME 10 mins
CHILL TIME 1 hour 15 mins

The gorgeous French opera cake is really just layers upon layers of deliciousness. In this recipe, an almond joconde sponge is joined by a luxurious dark-chocolate ganache, and a velvety, raspberry Swiss meringue buttercream that brings a tartness to balance out what is otherwise a rich dessert. The finishing touch is a glossy chocolate glaze, which, although sophisticated-looking, is easy as long as you use it within the temperature range for maximum shine.

ALMOND JOCONDE SPONGE
sunflower oil, for greasing
100g icing sugar
100g almond flour
3 eggs, at room temperature
3 egg whites, at room temperature
15g caster sugar
15g gluten-free flour blend
¼ teaspoon xanthan gum
fresh raspberries, to decorate

CHOCOLATE GANACHE
200g dark chocolate (60–70% cocoa solids), chopped
275g double cream

RASPBERRY SWISS MERINGUE BUTTERCREAM
50g freeze-dried raspberries ①
30g icing sugar
4 egg whites
250g caster sugar
¼ teaspoon cream of tartar
250g unsalted butter, softened

CHOCOLATE GLAZE
170g dark chocolate (about 60% cocoa solids), chopped
100g unsalted butter
1 tablespoon golden syrup

ALMOND JOCONDE SPONGE
+ Adjust the oven shelf to the middle position, preheat the oven to 180°C and line a 25 x 35cm shallow baking tray with baking paper. Lightly grease the baking paper with sunflower oil. ②
+ Sift together the icing sugar and almond flour. Add the whole eggs and, using a stand mixer with the whisk attachment or a hand mixer fitted with the double beaters, whisk on medium–high speed for about 5 minutes or until pale and doubled in volume.
+ In a separate bowl (with a clean whisk), whisk the egg whites until frothy, then slowly add the caster sugar, whisking continuously. Once you have added all the sugar, whisk for a further 2–3 minutes until the egg whites form stiff peaks.
+ Fold ⅓ of the egg whites into the almond mixture until fully combined. Sift in the gluten-free flour blend and xanthan gum, and fold them in to combine. Then, add the remaining egg whites and fold in gently, keeping the mixture as aerated as possible.
+ Pour the batter into the lined baking tray, smooth it out into an even layer, and bake for about 10–12 minutes or until golden brown on top and an inserted toothpick comes out clean.
+ Cool the sponge in the tray, then loosen the edges with a palette knife.

CHOCOLATE GANACHE
+ Place the chopped dark chocolate into a heatproof bowl.
+ In a saucepan, bring the double cream just to the boil and then pour it over the chocolate. Allow to stand for 4–5 minutes, then stir together into a smooth, glossy ganache.
+ Refrigerate the ganache for about 30 minutes, stirring every 10 minutes until it reaches a soft but spreadable consistency.

ASSEMBLING THE CAKE: PART I
+ Line a second shallow baking tray with baking paper. Invert the cooled sponge on to it and peel away the uppermost piece of baking paper.
+ Spread the ganache evenly across the sponge, from edge to edge. Refrigerate for 15–30 minutes until the ganache has set and feels firm to the touch. In the meantime, make the buttercream.

RASPBERRY SWISS MERINGUE BUTTERCREAM

+ In a food processor or blender, blitz the freeze-dried raspberries and icing sugar until fine. Pass the powder through a sieve to remove any larger pieces, and set aside until needed. ③
+ Mix the egg whites, caster sugar and cream of tartar in a heatproof bowl over a pan of simmering water. Stir continuously until the meringue mixture reaches 65°C and the sugar has dissolved.
+ Remove the mixture from the heat, transfer it to a stand mixer with the whisk attachment, or use a hand mixer fitted with the double beaters, and whisk on medium–high speed for 5–7 minutes until greatly increased in volume and at room temperature. It's very important that the meringue is at room temperature before you go to the next step.
+ Add the butter, 1–2 tablespoons at a time, while mixing continuously on medium speed. If using a stand mixer, I recommend switching to the paddle attachment for this step. ④
+ Continue until you've used up all the butter and the buttercream looks smooth and fluffy. Add the powdered raspberries and mix until evenly distributed.

ASSEMBLING THE CAKE: PART II

+ Spread the buttercream evenly across the chilled ganache, then refrigerate for 15–30 minutes until the buttercream has firmed up.
+ Trim away about 5mm–1cm edge to make a neat 23 x 30cm rectangle. Divide the rectangle into thirds, each measuring 10 x 23cm.
+ Stack the thirds one on top of the other, neaten the edges and sides if necessary, and refrigerate for at least 15 minutes.
+ To make the chocolate glaze, melt the chocolate, butter and golden syrup in a heatproof bowl over a pan of simmering water. Mix well until shiny and runny. Allow to cool until it reaches 32–34°C. ⑤
+ Transfer the chilled cake to a wire rack set over a baking sheet or large plate. Pour over the glaze, so that it covers both the top and the sides of the cake. If necessary, use an offset spatula to smooth out the glaze, but be careful not to disturb the underlying buttercream layer, as it can lead to white streaks in the glaze.
+ Allow to set for at least 30 minutes at room temperature, cut away the short ends to expose the layers, then cut into 6 equal slices (each will be just under 4cm wide). Decorate with fresh raspberries and serve.

STORAGE

3–4 days in a closed container in the fridge. Once refrigerated, bring up to room temperature for at least 10–15 minutes before serving.

PHOTOGRAPHY OVERLEAF ⟶

NOTES

① *Freeze-dried raspberries give the buttercream an intense colour and flavour without adding moisture (as a reduction or puréed fruit might do). This reduces the chances that the buttercream may split.*

② *Lightly greasing the baking paper prevents the sponge sticking to it (an otherwise common problem with relatively lean sponges such as this one).*

③ *Adding icing sugar to the freeze-dried raspberries makes it easier to process them to a fine powder and redistributes any trace of moisture in the raspberries to the sugar to help prevent the buttercream splitting.*

④ *Using the paddle attachment, rather than the whisk, incorporates less air into the buttercream at this point to give a smooth, velvety texture. For tips on troubleshooting Swiss meringue buttercream, see page 44.*

⑤ *At 32–34°C, the glaze sets with a beautiful, glossy finish and is runny enough to make a thin, delicate layer on top of the cake. At the same time, it's not so hot that it would melt the upper buttercream layer.*

CHOCOLATE + VANILLA MILLEFEUILLE

MAKES 5
PREP TIME 1 hour
(excluding the pastry)
CHILL TIME 35 mins
BAKE TIME 28 mins

You just can't go wrong with that classic combination of chocolate and vanilla, especially if it's combined with the flakiest layers of caramelised rough puff pastry. While millefeuille look terribly fancy, like something straight out of a French pâtisserie, they're actually surprisingly easy to make. Dusting the top with cocoa powder and icing sugar gives a professional-looking finish (with minimal effort) and ties together this classy, mouthwatering dessert.

1 quantity of **rough puff pastry** (see page 203)
1–2 tablespoons icing sugar, plus 1 tablespoon to decorate
1 tablespoon Dutch processed cocoa powder, to decorate

CHOCOLATE GANACHE
130g dark chocolate (60–70% cocoa solids), chopped
180g double cream

VANILLA WHIPPED CREAM
150g double cream, chilled
50g icing sugar, sifted, plus extra to taste if necessary
½ teaspoon vanilla bean paste

ROUGH PUFF PASTRY SHEETS
+ Pre-make the pastry according to the instructions on page 203.
+ Adjust the oven shelf to the middle position, preheat the oven to 200°C and remove the rough puff from the fridge or freezer to come up to room temperature as directed on page 203 (to prevent cracking).
+ On a lightly floured piece of baking paper, roll out the pastry to about 3mm thick and cut out a 22 x 44cm rectangle. Halve the rectangle to give two 22cm squares. Refrigerate for about 15 minutes until firm.
+ Slide the 2 squares of pastry with their baking paper on to a baking sheet, cover with another sheet of baking paper and top with a second baking sheet to prevent the pastry from puffing up too much in the oven. (If your baking sheet and oven can't accommodate baking the 2 pastry squares at once, you can bake them one after the other.)
+ Bake for 25–30 minutes or until the pastry is an even, deep golden brown. After about 20 minutes, start checking the pastry every 3–4 minutes so that it doesn't overbake.
+ Dust the rectangles with icing sugar, then return them, uncovered, to the oven for about 3–4 minutes for the sugar to caramelise. Keep an eye on them, as they can burn very easily.
+ Remove from the oven and transfer to a wire rack to cool.

CHOCOLATE GANACHE
+ Place the chopped dark chocolate in a heatproof bowl.
+ In a saucepan, bring the double cream just to the boil and pour it over the chocolate. Allow to stand for 4–5 minutes, then stir together into a smooth, glossy ganache.
+ Refrigerate the ganache for about 20 minutes, stirring every 5 minutes, until it reaches a pipeable consistency. (If you cool the ganache too much and it's too firm, just reheat it in the microwave or over a pan of simmering water until it reaches the right consistency.)

VANILLA WHIPPED CREAM

+ In a bowl, whip the double cream, icing sugar and vanilla paste until stiff peaks form. At this point, adjust the sweetness by adding more icing sugar, if you like.

ASSEMBLING THE MILLEFEUILLE

+ Using a sharp, serrated knife, trim the edges of the cooled pastry to give you 2 squares of 20cm. Divide each square into 8 smaller 5 x 10cm rectangles. You will need 15 of the 16 pastry rectangles for assembling the millefeuille.
+ Transfer the whipped cream and ganache each into a separate piping bag fitted with a large open star nozzle.
+ Pipe alternating mounds of cream and ganache on to 10 of the pastry rectangles – you should be able to fit 2 rows of 4 mounds on to each one. Make sure the order of cream and ganache is reversed for half of them and pair them up accordingly, placing one on top of the other. ①
+ Place a piece of paper diagonally across one of the remaining rectangles and dust the exposed half with icing sugar. Carefully flip the paper to cover the icing sugar and dust the other half with cocoa powder. Repeat for the remaining 4 rectangles.
+ Place each dusted rectangle on top of the stacked millefeuille, and serve immediately.

STORAGE

Best on the day, but you can keep the millefeuille in an airtight container in a cool, dry place for 2–3 days. If you keep them in the fridge, allow them to come up to room temperature for at least 30 minutes before serving (note that the ganache will be firmer because the chocolate crystallises in the fridge).

NOTE

① *Piping the whipped cream and ganache on top of each pastry rectangle before stacking them helps to keep all the layers an equal height (some people advise piping, stacking and piping, but this can compress the lower layers).*

PHOTOGRAPHY OVERLEAF ⟶

TIRAMISÙ

SERVES 10–12
PREP TIME 1 hour
BAKE TIME 2 x 20 mins
CHILL TIME 4 hours

LADYFINGERS
(MAKES 32–34)
3 eggs, separated
120g caster sugar
140g gluten-free flour blend
10g cornflour ①
1 teaspoon baking powder
¼ teaspoon xanthan gum
50g granulated sugar
30g icing sugar

MASCARPONE CREAM
750g mascarpone cheese, chilled
450g double cream, chilled
250g icing sugar
1–1½ teaspoons vanilla bean paste

TO ASSEMBLE
4–6 tablespoons instant coffee
    granules ②
2 tablespoons caster sugar
600g boiling water
1–2 tablespoons Dutch processed
    cocoa powder

Making your own ladyfinger biscuits may sound a bit scary, but is actually a piece of cake – and the element that takes this version of a classic Italian dessert from good to amazing. When making the ladyfingers, whisk the egg yolks and whites separately with sugar, for maximum aeration, and take care not to deflate the mixture too much. A dusting of icing sugar ensures the biscuits rise beautifully, and a slightly longer baking time at 170°C makes them crisp and perfect for soaking up lots of coffee. Make this dessert a day ahead if you can, to allow the flavours to mingle into something truly mind-blowing.

LADYFINGERS
+ Adjust the oven shelf to the middle position, preheat the oven to 170°C and line two baking sheets with baking paper.
+ Using a stand mixer with the whisk attachment or a hand mixer fitted with the double beaters, whisk the 3 egg whites and half the caster sugar together on high speed for 4–5 minutes until you get soft peaks. (Transfer the egg whites to a separate bowl, if necessary.)
+ Whisk the egg yolks and the remaining half of the caster sugar on high speed for 2–3 minutes until the mixture is thick enough to briefly mound up on itself as it falls off the whisk.
+ Gently fold the whisked yolks into the whites until only just combined (about 8 strokes with a spatula).
+ Sift the gluten-free flour blend, cornflour, baking powder and xanthan gum into the egg mixture and gently fold until no flour clumps remain (about 30–35 strokes).
+ Transfer the batter into a piping bag fitted with a 1cm round nozzle and pipe lines on to the lined baking sheets. The lines should be 10cm long and about 2cm wide, and spaced about 2cm apart to allow for expansion (you'll have about 16 ladyfingers per baking sheet).
+ Mix the granulated and icing sugars together, and generously dust half the ladyfingers with the mixture. (Dust the other baking sheet of ladyfingers just before baking them.) ③
+ One baking sheet at a time, bake the ladyfingers for 20–24 minutes until risen, slightly cracked and golden brown. Allow each batch to cool completely on the baking sheet.

MASCARPONE CREAM

+ Using a stand mixer with the whisk attachment or a hand mixer fitted with the double beaters, whip all the mascarpone cream ingredients together until they form very soft peaks. Set aside until needed.

ASSEMBLING THE TIRAMISÙ

+ Make up the coffee to your preferred strength using the coffee granules, sugar and boiling water. Allow to cool to warm or room temperature and pour into a wide bowl.
+ Prepare a 32 x 20 x 6cm (or thereabouts) baking or serving dish – it should be large enough to fit half the ladyfingers in a single layer.
+ Dip the ladyfingers into the coffee for 1–4 seconds until nicely soaked but not soggy. ④ Arrange them next to each other on the bottom of the dish as you go, until you've used up about half the biscuits. (If necessary, break the ladyfingers before dipping to fit them in nicely.)
+ Spoon half the mascarpone cream over the ladyfingers and smooth it out into an even layer.
+ Dip and arrange the remaining ladyfingers on top of the mascarpone cream, and then spoon the rest of the mascarpone cream on top, smoothing out the top.
+ Cover and refrigerate the tiramisù for at least 4 hours, or preferably overnight. Just before serving, dust the top with cocoa powder to finish.

STORAGE

2–3 days in a closed container in the fridge.

NOTES

① *Cornflour absorbs more moisture than a mixture with just gluten-free flour blend, thickening the batter and allowing you to pipe tall ladyfingers. (The fact that gluten-free flours anyway absorb more moisture than wheat flours actually makes gluten-free ladyfingers slightly easier to make than wheat-based ones.)*

② *The amount of coffee you add really depends on how strong you want the coffee flavour to be. I usually go with 6 tablespoons.*

③ *While the granulated sugar is there for the crunch, the icing sugar is absolutely paramount for the characteristic tall shape of the ladyfingers. Without it, the ladyfingers will spread out significantly during baking and rise only minimally.*

④ *The dipping time determines the intensity of the coffee flavour and the soft, melt-in-the-mouth-ness of your ladyfingers. Adjust the dipping time according to the temperature of the coffee – if it's warm, dip for a shorter time (1–2 seconds); if it's completely cold, dip for longer (3–4 seconds).*

PHOTOGRAPHY OVERLEAF ⟶

BACI DI DAMA

MAKES 36
PREP TIME 45 mins
COOK TIME 10 mins
BAKE TIME 2 x 12 mins

Toasted hazelnuts and chocolate, in crumbly, melt-in-the-mouth, one-bite cookie form – surely, this is cookie heaven. They might be tiny, but baci di dama concentrate an incredible amount of flavour into every mouthful. Plus, they will make your kitchen smell like the most amazing Italian pasticceria, packed with rows and rows of tempting, beautifully displayed treats.

130g hazelnuts ①
100g caster sugar
100g unsalted butter, softened
130g gluten-free flour blend
½ teaspoon xanthan gum
¼ teaspoon salt
85g dark chocolate (60–70% cocoa solids), melted

NOTE
① *You can vary the amounts of hazelnuts and gluten-free flour blend from 125g–135g (keeping a 1:1 ratio), but 130g gives a dough that is both easy to work with and keeps its shape during baking. Dough made with 125g of each will spread and crack more, but is easier to roll; with 135g, the dough is much more crumbly and difficult to roll, but gives a neater shape with less cracking.*

+ Toast the hazelnuts in a frying pan over a medium–high heat for about 10 minutes until they have deep golden-brown spots and their skins start loosening. Transfer them into a clean tea towel and rub them vigorously to remove the skins. Allow to cool.
+ Adjust the oven shelf to the middle position and preheat the oven to 160°C and line two baking sheets with baking paper.
+ Blitz the toasted hazelnuts in a food processor until fine (about 30–60 seconds). Add the caster sugar and process for a further 30–60 seconds.
+ Add the butter and process until smooth and creamy.
+ Add the gluten-free flour blend, xanthan gum and salt, and pulse until combined to a crumbly dough. Turn out the dough on to a work surface and knead it briefly to bring it into a ball.
+ Roll teaspoon-sized dough portions (each about 6g) into small balls (you'll need 72 to make 36 completed baci di dama) and place them on the lined baking sheets about 1cm apart. The dough might be slightly crumbly, but should form small balls without major problems.
+ One tray at a time, bake the cookies for about 12 minutes until they just start turning a light golden brown. Allow the cookies to cool completely on the baking sheets, as they are very soft and crumbly when warm.
+ Put a small dollop of chocolate on the flat base of 1 cookie (about 1.5g, the size of a large chocolate chip) and sandwich it together with another cookie, flat sides facing. Repeat for all the cookies so that you end up with 36 baci di dama. Allow the chocolate to set at room temperature before serving.

STORAGE
1 week in a closed container at room temperature.

SMOOTH + CREAMY NEW YORK-STYLE CHEESECAKE

SERVES 10–12
PREP TIME 45 mins
BAKE TIME 1 hour
CHILL TIME 4 hours

Who doesn't love a slice of vanilla-scented, creamy New York cheesecake with a buttery, crumbly, crushed-cookie crust? This recipe requires no water bath for baking, because room-temperature ingredients, mixing the filling on the lowest speed (or doing it by hand), baking at a low 140°C, and allowing the baked cheesecake to cool to room temperature in a turned-off oven all contribute to perfect results – a smooth, velvety texture with no cracking in sight.

½ quantity of **digestive biscuits**
  (16 biscuits; see page 162)
40g unsalted butter, melted ①
800g full-fat cream cheese,
  at room temperature
150g full-fat plain yoghurt,
  at room temperature
200g caster sugar
20g cornflour
4 eggs, at room temperature
1 tablespoon lemon juice
1 teaspoon vanilla bean paste
zest of 1 unwaxed lemon (optional)
tinned pitted cherries in syrup,
  to serve (optional)

NOTES

① *This butter quantity assumes you're using the digestives on page 162. If not, you may need to increase the butter. Add it slowly until you reach the wet-sand texture and the mixture holds its shape when pressed.*

② *The baking time of 50 minutes is a guide. Start checking the cheesecake about 10 minutes earlier – look at the wobble in the centre and the puffiness at the edges to decide if it's ready.*

③ *The texture and flavour of the cheesecake are better the next day. Bake a day ahead if you can.*

+ Adjust the oven shelf to the middle position, preheat the oven to 180°C and line the bottom and sides of a 20cm round springform tin with baking paper.
+ Use a food processor to crush the digestive biscuits to fine crumbs, or put them in a sturdy zip-lock bag and use a rolling pin to bash them.
+ Weigh out 200g of the crushed biscuits and mix them with the melted butter until the texture resembles wet sand. Transfer to the lined tin and, using the flat base of a glass or measuring cup, compress to an even layer with a 1.5–2cm raised rim around the edge.
+ Place the tin on a baking sheet or tray to catch any leaks, and bake the base for about 10 minutes, then remove from the oven and allow to cool until warm. Reduce the oven temperature to 140°C.
+ Using a stand mixer with the paddle attachment on the lowest speed, or by hand using a balloon whisk (avoid an electric hand mixer, which will incorporate too much air, causing the cheesecake to crack), mix together the cream cheese and yoghurt until smooth. Take great care not to aerate the ingredients too much.
+ Add the caster sugar and cornflour and combine. Add the eggs, one at a time, mixing well after each addition, then add the lemon juice, vanilla paste and the lemon zest (if using), and mix until evenly incorporated.
+ Pour the cheesecake filling over the base, smooth out the top, and bake for about 50 minutes. ② When the cheesecake is done, the edges will be slightly puffed up and may have some small, thin cracks (these will close up as the sides settle down on cooling). There should be about a 9cm-wide circular dip in the middle that will wobble as you gently shake the pan. The top of the cake should be a creamy off-white colour with a hint of light gold around the edges.
+ Turn off the oven, leave the oven door slightly ajar and cool the cheesecake inside for about 1 hour until it's at room temperature.
+ Refrigerate the cheesecake for at least 4 hours or preferably overnight, then remove from the tin. ③ Serve topped with cherries, if you wish.

STORAGE
3–4 days in a closed container in the fridge.

MIXED BERRY COBBLER

SERVES 6
PREP TIME 30 mins
COOK TIME 5 mins
BAKE TIME 45 mins

BERRY FILLING

700g fresh or frozen mixed berries,
    such as raspberries, blueberries
    and blackberries
100g caster sugar
1 tablespoon cornflour
1 teaspoon vanilla bean paste

BISCUIT TOPPING

120g gluten-free flour blend
50g caster sugar
1½ teaspoons baking powder
½ teaspoon xanthan gum
¼ teaspoon salt
180g double cream, plus
    1 tablespoon for glazing
1 tablespoon granulated sugar
ice cream or whipped double
    cream, to serve (optional)

NOTES

① *Always thaw frozen berries first,
    otherwise the filling will have
    to thaw in the oven, giving a
    longer baking time (over 1 hour
    15 minutes) and so more time for
    the biscuit topping to absorb the
    berry juices and become soggy.*

② *If you want the biscuit topping to
    be flakier, you can 'laminate' the
    dough: roll it out, do a letter fold,
    then an optional simple fold in
    half – as for the scones (see page
    327), just without brushing the
    dough with melted butter.*

There are only a few rules when it comes to making this traditional American dessert. You can use pretty much any berry you like – although if you're using only one type (rather than a mixture), I recommend blueberries or blackberries over strawberries or raspberries, as the latter give a far more watery filling. A mixture of several berries gives the richest flavour, of all. Use fresh or frozen berries, but, unless you want to end up with a sad, soggy topping, always thaw out frozen fruit and reduce the released juices on the hob before returning them to the filling. Although this adds an extra step, it makes the world of difference to the results.

BERRY FILLING

+   Adjust the oven shelf to the middle position and preheat the oven to 200°C. On your work surface, place a 15 x 23cm baking dish (at least 5cm deep) on a baking sheet (to catch any juices during baking).
+   If using frozen fruit: thaw the berries completely and drain the released juices into a saucepan. Cook the berry juices over a medium–high heat, stirring frequently, for about 5 minutes until noticeably thickened. Mix the reduced juices back into the berries. ①
+   Mix all the berry filling ingredients together and transfer them into the baking dish.

BISCUIT TOPPING

+   Whisk the gluten-free flour blend, sugar, baking powder, xanthan gum and salt together in a bowl. Add the double cream and mix everything together into a smooth dough. Knead it briefly to form a ball. ②
+   Transfer the dough to a lightly floured surface and roll it out until about 1cm thick. Using a 4.5cm round cutter, cut out the biscuits, re-rolling the trimmings as necessary to give 15–17 biscuits.
+   Arrange the biscuits on top of the berry filling as closely together as possible. Brush with double cream and sprinkle with granulated sugar.
+   Bake for 45–55 minutes or until the biscuit topping is a deep golden brown and a toothpick inserted into the topping comes out clean, with no unbaked dough attached. If the topping browns too quickly, cover with foil (shiny side up) and bake until done.
+   Allow to cool for 10 minutes, then serve warm – with ice cream or whipped cream, if you wish.

STORAGE

Best on the day. Or, cover the baked cobbler with cling film and store for up to 2 days. Reheat in the microwave for 15–30 seconds to serve.

LINZER COOKIES

MAKES 30
PREP TIME 45 mins
CHILL TIME 15 mins
BAKE TIME 2 x 12 mins

375g gluten-free flour blend
200g finely ground almonds
200g caster sugar
½ teaspoon xanthan gum
¼ teaspoon salt
200g unsalted butter, softened
1 egg, at room temperature
1 egg yolk, at room temperature
½ teaspoon vanilla bean paste
200g raspberry or strawberry jam
icing sugar, for dusting

NOTES

① *As you want the cut-out cookies to keep their shape, and neat, straight edges, kneading the cookie dough is preferable to creaming the butter and sugar together with a stand mixer. The latter aerates the dough, and the trapped air acts similarly to chemical leavening agents, causing the cookies to spread in the oven. (Note the absence of baking powder and bicarbonate of soda for the same reason.)*

② *Chilling the cookie dough before cutting out the holes ensures the cookies keep their shape – room-temperature cookies sometimes become distorted as you cut.*

These jam-filled, icing-sugar-dusted, crumbly, melt-in-the-mouth classic Austrian cookies never fail to put a smile on my face. And while they're delicious on the day they're baked, the cookies truly come into their own a few days later, after they have taken on some of the moisture from the jam and the flavours have had time to mingle. I prefer finely ground almonds (with their skins on, not blanched) to almond flour because they carry more flavour. Almond flour is great when you want a subtle or even unnoticeable almond flavour – in linzer cookies the almonds really need to shine.

+ Adjust the oven shelf to the middle position, preheat the oven to 180°C and line two baking sheets with baking paper.
+ In a large bowl, whisk together the gluten-free flour blend, ground almonds, sugar, xanthan gum and salt. Add the butter, whole egg, egg yolk and vanilla paste, and knead everything together to a smooth (but not sticky) dough. ① You should be able to roll out the dough and cut out the cookies easily, but if you think the dough is too soft, refrigerate it for 10–15 minutes before continuing.
+ Roll out the cookie dough on a generously floured surface until about 3mm thick. Dust the top of the cookie dough or sprinkle your rolling pin with flour as needed.
+ Using a 6cm round cutter, cut out 60 dough discs, re-rolling the trimmings as necessary. Transfer the discs to the lined baking sheets (30 on each sheet) and refrigerate for 15–30 minutes.
+ Once the unbaked cookies are chilled and firm to the touch, use a small cookie cutter (I used a heart-shaped one, but any works) to stamp out holes in 30 of them (use all from one tray, so that you can cook those with a hole and those without using slightly different baking times). ② (Gather up and re-roll the heart-shaped, or other cut-outs to make more cookies, or just bake the shapes as mini-bites.)
+ One tray at a time, bake the whole cookies (no holes) for 12–14 minutes and then the cut-out cookies for 10–12 minutes or until each type is very light golden around the edges.
+ Allow to cool on the baking sheets for about 10 minutes, then transfer all the cookies to a wire rack to cool completely.
+ To assemble, spread about 1 teaspoon of raspberry or strawberry jam on the base of the cookies without the holes. Dust the holey (top) cookies with icing sugar, and sandwich the tops and bottoms together to give 30 complete linzer cookies.

STORAGE
1 week in a closed container at room temperature.

SUNKEN APPLE CAKE

SERVES 10–12
PREP TIME 45 mins
BAKE TIME 30 mins

A German sunken apple cake certainly is eye-catching, with the prettily arranged sliced apples, a golden sponge and a shiny glaze of apricot jam. The tart apples nicely balance the sweetness of the cinnamon-y sponge, creating a simple cake that earns pride of place at any afternoon tea.

2 sweet–tart, firm apples (such as Granny Smith, Pink Lady or Braeburn), peeled, cored and quartered
2–3 tablespoons lemon juice
115g unsalted butter, softened
60g light brown soft sugar
60g caster sugar
½ teaspoon vanilla bean paste
3 eggs, at room temperature
150g gluten-free flour blend
30g almond flour
2 teaspoons baking powder
½ teaspoon xanthan gum
½–1 teaspoon ground cinnamon, plus extra to sprinkle
¼ teaspoon salt
100g whole milk, at room temperature
2–3 tablespoons apricot jam

NOTE

① *Adding the dry ingredients and milk in alternating batches maintains the emulsion created by the butter, sugar and eggs, and gives a perfectly fluffy cake with a delicate crumb.*

+ Adjust the oven shelf to the middle position, preheat the oven to 180°C and line a 20cm round springform tin with baking paper.
+ Thinly slice each apple quarter lengthways, being careful not to cut all the way through to the core so that the quarters remain intact. Place the quarters on a plate and drizzle over the lemon juice (to prevent oxidation and browning) and set aside until needed.
+ Using a stand mixer with the paddle attachment or a hand mixer fitted with the double beaters, cream the butter and both sugars together until pale and fluffy. Add the vanilla paste and mix until combined.
+ Add the eggs, one at a time, mixing well after each addition, until well combined.
+ In a separate bowl, sift together the gluten-free flour blend, almond flour, baking powder, xanthan gum, cinnamon and salt.
+ Beginning and ending with the dry ingredients, alternately add the dry ingredients (in three batches) and the milk (in two batches) to the butter mixture, mixing well after each addition, until you get a smooth batter with no flour clumps. ①
+ Transfer the batter to the lined springform tin, smooth out the top and arrange the sliced apple quarters on top. (Discard any remaining lemon juice.)
+ Bake for 30–35 minutes or until the cake is golden brown on top and an inserted toothpick comes out clean. If the cake starts browning too quickly, cover with foil (shiny side up) and bake until done.
+ Remove the sponge from the oven, allow to cool in the baking tin for about 10–15 minutes, then transfer to a wire rack to cool completely.
+ Heat the apricot jam on the hob or in the microwave until runny, then brush it all over the top of the cake. Sprinkle with cinnamon and serve.

STORAGE
3–4 days in a closed container at room temperature.

LAMINGTONS

MAKES 16
PREP TIME 1 hour
BAKE TIME 35 mins
COOK TIME 5 mins
CHILL TIME 15 mins

Originating from Australia, with a buttery sponge, a layer of tart raspberry jam, the most amazing chocolate icing and a coconut coating, Lamingtons have charm galore. Things always get a bit messy when making them – hands covered in chocolate, the kitchen covered in desiccated coconut… but I think that's part of their appeal. I like my Lamingtons on the larger side (these are 5cm cubes), but you can make them smaller and bite-sized, if you prefer.

VANILLA SPONGE

150g unsalted butter, softened
180g caster sugar
1 teaspoon vanilla bean paste
3 eggs, at room temperature
180g gluten-free flour blend
½ teaspoon xanthan gum
1½ teaspoons baking powder
¼ teaspoon salt
120g whole milk, at room
    temperature

CHOCOLATE ICING

200g dark chocolate (60–70%
    cocoa solids), chopped
120g icing sugar
100g whole milk
10g Dutch processed cocoa
    powder
10g unsalted butter

TO ASSEMBLE

120g raspberry or strawberry jam
200g desiccated coconut

NOTES

① *Adding the dry ingredients
    and milk in alternating batches
    maintains the emulsion created
    by the butter, sugar and eggs,
    and gives a perfectly fluffy
    sponge with a delicate crumb.*

② *Hand-dipping the Lamingtons is
    the best way to keep the squares
    intact and cover them fully.*

VANILLA SPONGE

+ Adjust the oven shelf to the middle position, preheat the oven to 180°C and line a 20cm square baking tin with baking paper.
+ Using a stand mixer with the paddle attachment or a hand mixer fitted with the double beaters, cream together the butter, sugar and vanilla paste on a medium–high speed for about 5 minutes until pale and fluffy.
+ Incorporate the eggs, one at a time, mixing well after each addition.
+ Sift together the gluten-free flour blend, xanthan gum, baking powder and salt. Alternately add the dry ingredients (in thirds) and the milk (half at a time) to the butter mixture. ① Mix to a smooth batter.
+ Transfer the batter to the lined baking tin and smooth out the top. Bake for about 35–40 minutes or until an inserted toothpick comes out clean.
+ Remove the sponge from the oven, allow to cool in the baking tin for about 10 minutes, then turn out on to a wire rack to cool completely.

CHOCOLATE ICING

+ Combine all the icing ingredients in a saucepan over a medium–high heat, stirring continuously, until all the chocolate has melted and you have a smooth, glossy icing (about 5 minutes). Set aside until needed.

ASSEMBLING THE LAMINGTONS

+ Cut the cooled sponge in half horizontally, and spread the jam evenly across the bottom half. Cover with the top sponge, pressing down slightly, and let the cake stand for 20–30 minutes so that the sponges absorb the jam a little, which will help stick them tighter together.
+ Cut the sponge into 16 squares of 5cm.
+ If the chocolate icing has firmed up too much, reheat it until it becomes runny again. (You can add 1–2 tablespoons milk to thin it out slightly, if necessary.) Dip each sponge square into the icing, and place it on to a wire rack for 2–3 minutes, allowing any excess icing to drip away. ② Then, roll it in desiccated coconut until fully covered.
+ Leave the Lamingtons at room temperature for at least 30 minutes (or in the fridge for at least 15 minutes) until the icing sets, then serve.

STORAGE
3–4 days in a closed container at room temperature.

CONVERSION TABLES

These tables give US and imperial measurements vs metric equivalents. Note that, for volume, your measuring cups and spoons might not be correctly calibrated, which can affect your bakes. (For example, the volume of one of my '1 US cup' measures is 225ml, rather than the correct 240ml.)

When measuring gluten-free flours and blends, note that flours from different brands may have different densities, and therefore a different weight for the same volume. Finally, when measuring flour by volume, make sure you do so correctly: 'fluff up' the flour within the container using a spoon or a fork, then spoon the flour into your measuring cup, and level it off with the back of a knife or other straight-edged utensil.

Above all, if possible: skip the volume measurements altogether – use digital scales and weigh all your ingredients.

Confusingly, some ingredients have different names in the US and UK. I've included a list of these on page 374.

US volume conversions

ML	US VOLUME MEASURE
240	1 cup
180	¾ cup
160	⅔ cup
120	½ cup
80	⅓ cup
60	¼ cup
30	⅛ cup
15	1 tablespoon
7.5	½ tablespoon
5	1 teaspoon
2.5	½ teaspoon
1.25	¼ teaspoon

Imperial volume conversions

ML	IMPERIAL VOLUME MEASURE
250	1 cup
185	¾ cup
165	⅔ cup
125	½ cup
85	⅓ cup
60	¼ cup
30	⅛ cup
15	1 tablespoon
7.5	½ tablespoon
5	1 teaspoon
2.5	½ teaspoon
1.25	¼ teaspoon

Oven temperatures

°C	°C FAN	°F	GAS MARK
100	80	200	½
110	90	225	½
130	110	250	1
140	120	275	1
150	130	300	2
160	140	315	2–3
170	150	325	3
180	160	350	4
190	170	375	5
200	180	400	6
210	190	415	6–7
220	200	425	7
230	210	450	8
240	220	475	8

Gluten-free flours + blends: weight (grammes) vs US volume measurements

GLUTEN-FREE FLOUR OR BLEND	G PER 1 US CUP	G PER 1 US TABLESPOON
almond flour	100	6
arrowroot starch	120	7.5
buckwheat flour	150	9.5
cornflour	115	7
maize flour	125	8
millet flour	135	8.5
oat flour	90	5.5
potato starch	165	10.5
quinoa flour	120	7.5
rice flour, brown	160	10
rice flour, white	175	11
sorghum flour	130	8
tapioca starch	115	7
teff flour, white	155	9.5
Doves Farm Freee gluten-free flour (plain)	130	8
store-brand gluten-free flour blends	~130	~8
DIY blend 1	150	9.5
DIY blend 2	140	9

Dairy products: weight (grammes) vs US volume measurements

DAIRY PRODUCT	G PER 1 US CUP	G PER 1 US TABLESPOON
butter, unsalted	226 (2 sticks)	14
buttermilk	240	15
cheddar, coarsely grated	110	6
cream cheese	225	14
double cream	230	14.5
mascarpone cheese	230	14.5
parmesan, finely grated	80	5
soured cream	230	14.5
yoghurt, Greek	230	14.5
yoghurt, plain	230	14.5

Non-dairy fats + oils: weight (grammes) vs US volume measurements

NON-DAIRY FATS + OILS	G PER 1 US CUP	G PER 1 US TABLESPOON
coconut oil, melted	220	14
coconut oil, softened	220	14
olive oil	215	13.5
sunflower oil	215	13.5

Solid/crystalline sweeteners: weight (grammes) vs US volume measurements

SOLID/CRYSTALLINE SWEETENER	G PER 1 US CUP	G PER 1 US TABLESPOON
caster sugar	215	13.5
granulated sugar	200	12.5
icing sugar	120	7.5
soft brown sugar, light or dark	200 (packed)	12.5

Liquid sweeteners: weight (grammes) vs US volume measurements

LIQUID SWEETENER	G PER 1 US CUP	G PER 1 US TABLESPOON
golden syrup	350	22
honey, runny	350	22
maple syrup	315	20
molasses	325	20.5

Nuts + seeds (including butters + pastes): weight (grammes) vs US volume measurements

NUTS + SEEDS	G PER 1 US CUP	G PER 1 US TABLESPOON
almonds, whole	145	/
almonds, flaked	90	5.5
hazelnuts, whole	140	/
mixed seeds	160	10
peanut butter	270	17
peanuts, whole	140	/
pecans, whole	120	/
poppy seeds	/	10
sesame seeds	/	10
tahini paste	245	15.5
walnuts, whole	125	/

Fruit: weight (grammes) vs US volume measurements

FRUIT	G PER 1 US CUP	G PER 1 US TABLESPOON
blackberries, fresh or frozen	140	/
blueberries, fresh or frozen	130	/
raspberries, fresh or frozen	120	/
raspberries, freeze-dried	30	2
strawberries, fresh (whole)	130	/
strawberries, freeze-dried	30	2

Other important ingredients: weight (grammes) vs US volume measurements

INGREDIENT	G PER 1 US CUP	G PER 1 US TABLESPOON
baking powder	/	12
bicarbonate of soda	/	18
chocolate, chopped	~180	~11
chocolate chips	180	11
cocoa powder	100	6
desiccated coconut	80	5
egg white powder	/	6.5
gelatine powder	/	9
jam/conserve	315	20
oats, whole	100	6
oats, ground	110	7
psyllium husk	80	5
salt	/	17.5
sultanas/raisins	145	9
xanthan gum	/	8.5
yeast	/	12

Length conversions

MM/CM	INCHES	CM	INCHES
2mm	¹⁄₁₆ in	19cm	7½ in
3mm	⅛ in	20cm	8 in
5mm	¼ in	21cm	8¼ in
8mm	⅜ in	22cm	8½ in
1cm	½ in	23cm	9 in
1.5cm	⅝ in	24cm	9½ in
2cm	¾ in	25cm	10 in
2.cm	1 in	26cm	10½ in
3cm	1¼ in	27cm	10¾ in
4cm	1½ in	28cm	11¼ in
4.5cm	1¾ in	29cm	11½ in
5cm	2 in	30cm	12 in
5.5cm	2¼ in	31cm	12½ in
6cm	2½ in	33cm	13 in
7cm	2¾ in	34cm	13½ in
8cm	3¼ in	35cm	14 in
9cm	3½ in	37cm	14½ in
9.5cm	3¾ in	38cm	15 in
10cm	4 in	39cm	15½ in
11cm	4¼ in	40cm	16 in
12cm	4½ in	42cm	16½ in
13cm	5 in	43cm	17 in
14cm	5½ in	44cm	17½ in
15cm	6 in	46cm	18 in
16cm	6¼ in	48cm	19 in
17cm	6½ in	50cm	20 in
18cm	7 in		

Glossary of terms

UK TERMS	US TERMS
Bicarbonate of soda	Baking soda
Caster sugar	Superfine sugar
Cornflour	Corn starch
Double cream	Heavy cream
Golden syrup	There is no ideal substitute, so it is worth seeking out
Icing sugar	Powdered or confectioner's sugar
Light brown soft sugar	Light brown sugar
Maize flour	Corn flour (finely milled cornmeal)

INDEX

raisins
  dried fruit soda bread 316
raspberries
  mascarpone raspberry frosting 66–7
  pink drizzle cake 61
  raspberry + chocolate opera cake
    346–7
  raspberry frangipane tart 239
  raspberry Swiss meringue buttercream
    346–7
  raspberry traybake 66–7
  raspberry whipped cream 70–1
  raspberry white chocolate brownies
    134
  vanilla + raspberry Swiss roll 70–1
raspberry jam
  Lamingtons 369
  Linzer cookies 365
  peanut butter + jam thumbprint
    cookies 165
razors, scoring 30
reverse creaming method 38–9, 38
rice flour
  brown rice flour 15, 16, 17, 20, 254
  white rice flour 15, 16, 17
rising, bread-making 256–8, 256–7
rolling pins 28
rolls see buns and rolls
rosemary crackers 174
rosemary focaccia 298
rotating cake stands 28
rough puff pastry 193, 193, 202, 203
  oven temperatures 198
royal icing 173

S
salt 22
  in cookies 151
salted-caramel-stuffed brownies 136–7
salted-caramel-stuffed chocolate chip
  cookie bars 178–9
sandwich bread 276–7
scales, digital 25
scaling recipes up and down 42–3, 43
scones, extra-flaky 327
scoops, ice-cream 27

scoring bread dough 260
scoring razors 30
scrapers 29
seeded buns 284
sesame seeds
  tahini cookies 166
shiny top brownies 124–5, 126
shortbread
  lemon bars 188
  shortbread biscuits 161
shortcrust pastry
  chocolate sweet shortcrust pastry 207
  oven temperatures 198
  percentage of flour 34
  sweet shortcrust pastry 194, 194, 204,
    205
sieves 27
silicone mats 27
simple vanilla cake 50–1
skewers 29
skillets 29
  peanut-butter-stuffed chocolate chunk
    skillet cookie 177
small cakes
  chocolate lava cakes 78
  see also bars; brownies; cupcakes
s'mores bars 184–5
s'mores cupcakes 89
soda bread, dried fruit 316
softened butter 21
sorghum flour 15, 17, 20, 254
spatulas 26, 27
sponge
  almond joconde sponge 346
  chocolate Swiss roll 74–5
  hazelnut + milk chocolate cake 54–5
  Lamingtons 369
  simple vanilla cake 50–1
  super-moist chocolate cake 46–7
  vanilla Swiss roll 70–1
  Victoria sponge cake 328
spoons, wooden 26
squash see butternut squash
stand mixers 26, 255
standard creaming method 37–8, 38
starch gluten-free flours 15–17

bread-making 253–4
steam ovens, bread-making 261
storage, bread 264
strawberries
  chocolate-glazed strawberry baked
    doughnuts 331
  strawberries + cream tart 240
  strawberry + white chocolate cupcakes
    108
  strawberry buttercream 108
  strawberry lemonade tartlets 232
strawberry jam
  Lamingtons 369
  Linzer cookies 365
  Victoria sponge cake 328
substitutions, gluten-free flour 20, 254
sugar 22
  in brownies 123
  in cookies 148–9
  creaming methods 37–9, 38
  dissolving in butter myth 125
  simple sugar icing 312–13
  sugar syrup glaze 308–9
sugar cookies
  chocolate sugar cookies 158
  no-fuss, no-chill, no-spread one-bowl
    sugar cookies 158
sultanas
  dried fruit soda bread 316
sunken apple cake 366
super-moist chocolate cake 46–7
sweet shortcrust pastry 194, 194, 204,
    205
  chocolate sweet shortcrust pastry 207
Swiss meringue
  hot chocolate tart 244–5
  lemon meringue pie 216–17
  s'mores cupcakes 89
  toasted marshmallow brownies 142
Swiss meringue buttercream 44–5
  poppy seed 104–5
  raspberry 346–7
  vanilla 50, 92–3
Swiss roll
  chocolate Swiss roll 74–5
  vanilla + raspberry Swiss roll 70–1

ACKNOWLEDGEMENTS

Creating this book was an adventure, a challenge and a privilege –
and it would've been impossible without the help of so many
wonderful people, who lent me their time, knowledge, expertise
and support along the way.

First, I have to thank the readers and followers of The Loopy Whisk,
without whom this book would never have happened. When I published
my first post way back in 2016, I never could've imagined it would have
such a profound effect on my life, let alone that it would result in me
writing a book. Thank you for reading my words, making my recipes
and sharing my excitement for all things chocolate.

To my wonderful agent Tessa David: thank you for sending that
first email, back in September 2018, with the subject reading only
'Cookbook' (which almost got deleted, because I was 100% convinced
it was spam... thankfully, I didn't press the 'delete' button). Your advice
and calming voice were indispensable in navigating the previously
unfamiliar world of publishing.

Thank you to the amazing Bloomsbury team. Somehow, I had
the privilege of working with a total of four amazing editors throughout
the creation of this book, and the end result is all the better for it.
To Natalie Bellos, my first editor: thank you for pushing the science to
the forefront and encouraging my inner nerd to roam free – while also
being there to point out that, just maybe, people don't need to know
the exact molecular structure of xanthan gum and that sometimes,
simple is better than a half-page formula derivation. To Lisa Pendreigh,
for helping give shape to my confused ideas about what the ideal
cover should be. To Rowan Yapp, for all the support in getting the
book across the finishing line in all its shiny, polished glory. And to
Kitty Stogdon, who was there from start to finish, and answered all
my numerous questions and worries and doubts with an endless
amount of patience.

I had the good fortune of working with Judy Barratt as my copy
editor, who is about as nitpicky as me (and I mean that as the highest
of compliments), and managed the seemingly impossible task of
condensing my frequently long-winded sciency ramblings into a
much more coherent, engaging and easier-to-follow book.

Thank you to Peter Dawson from Grade Design, who somehow
made a text-and-information-heavy book look fun and approachable –
no daunting pages-long blocks of text in sight. Beyond the gorgeous
book interior, the road to the current (amazing, glorious) cover was long
and at times seemed hopeless – thank you for persevering and creating
a cover that combines all of the lusciousness of baking with the
sciency precision of it.

This book was mostly written in the evenings and at the weekends, the weekdays being spent in the research lab, working on my PhD project. It was exhausting, but coming to the lab was never a chore thanks to my wonderful colleagues in the O'Hare research group (who bravely polished off all the cakes and cookies I would bring in on Monday, after a long weekend of baking). I am still convinced that only a crazy person would attempt to finish their PhD while writing a book, and I couldn't have managed it without the guidance of Prof Dermot O'Hare and Dr JC Buffet. The biggest thank you to (in alphabetical order, lest anyone starts complaining) Dana, Jess, Joyce, Justin, Phil and Tom, for indulging my baking-related monologues and making me laugh even when completely stressed out.

I'm the person who leaves the best (and most important) part for last. I believe that the last bite of a dish or dessert should comprise only the very best bits, for an unforgettable finish to a wonderful experience. In that spirit, I leave the most important thanks for last.

Thank you to my parents, who taught me that I can achieve anything I set my mind to. My mum, who gave me a love of all things cooking and baking amid endless laughter (though how I ended up writing a book of recipes when she never, ever follows them nobody knows) and my dad, who started my enthusiasm for all things science with discussions about stars and dinosaurs and gravity, and who guides me on my photography journey (even when I'm a bit too stubborn to listen). None of this would be possible without you. I love you to the moon and back.

ABOUT THE AUTHOR

Katarina Cermelj is a food writer, photographer and creator of the popular baking blog, The Loopy Whisk, where she shares her sumptuous allergy-friendly recipes. She has a PhD in Inorganic Chemistry from the University of Oxford, having previously completed her undergraduate studies in Chemistry at the same university. Her scientific background, a love of research and a good deal of stubbornness allowed her to develop mouthwatering recipes suitable for a wide range of dietary restrictions, including gluten-, dairy- and egg-free. Originally from Slovenia, Katarina now lives near Oxford. *Baked to Perfection* is her first book.

theloopywhisk.com   🅕 🅣 🅘 @theloopywhisk